AF845310

Innovative Alternatives to Minimize Arsenic, Perchlorate, and Nitrate Residuals

About the Awwa Research Foundation

Awwa Research Foundation

The Awwa Research Foundation (AwwaRF) is a member-supported, international, nonprofit organization that sponsors research to enable water utilities, public health agencies, and other professionals to provide safe and affordable drinking water to consumers.

The Foundation's mission is to advance the science of water to improve the quality of life. To achieve this mission, the Foundation sponsors studies on all aspects of drinking water, including supply and resources, treatment, monitoring and analysis, distribution, management, and health effects. Funding for research is provided primarily by subscription payments from approximately 1,000 utilities, consulting firms, and manufacturers in North America and abroad. Additional funding comes from collaborative partnerships with other national and international organizations, allowing for resources to be leveraged, expertise to be shared, and broad-based knowledge to be developed and disseminated. Government funding serves as a third source of research dollars.

From its headquarters in Denver, Colorado, the Foundation's staff directs and supports the efforts of more than 800 volunteers who serve on the board of trustees and various committees. These volunteers represent many facets of the water industry, and contribute their expertise to select and monitor research studies that benefit the entire drinking water community.

The results of research are disseminated through a number of channels, including reports, the Web site, conferences, and periodicals.

For subscribers, the Foundation serves as a cooperative program in which water suppliers unite to pool their resources. By applying Foundation research findings, these water suppliers can save substantial costs and stay on the leading edge of drinking water science and technology. Since its inception, AwwaRF has supplied the water community with more than $300 million in applied research.

More information about the Foundation and how to become a subscriber is available on the Web at **www.awwarf.org**.

Innovative Alternatives to Minimize Arsenic, Perchlorate, and Nitrate Residuals

Prepared by:
Joon H. Min, **Lina Boulos**, and **Jess C. Brown**
Carollo Engineers, P.C., Fountain Valley, CA 92708

David A. Cornwell and **Yann Le Gouellec**
Environmental Engineering & Technology, Newport News, VA 23606

Edward N. Coppola, **J. Steven Baxley**, and **Jeffrey A. Rine**
Applied Research Associates, Panama City, FL 32401

and

Janet G. Hering and **Nermin Vural**
California Institute of Technology, Pasadena, CA 91125

Jointly sponsored by:
Awwa Research Foundation
6666 West Quincy Avenue, Denver, CO 80235-3098

United States Environmental Protection Agency
Washington D.C.

and

Association of California Water Agencies
Sacramento, CA

Published by:

Awwa Research Foundation
Advancing the Science of Water®

DISCLAIMER

CONTENTS

TABLES

FIGURES

FOREWORD

The Awwa Research Foundation (AwwaRF) is a nonprofit corporation that is dedicated to the implementation of a research effort to help utilities respond to regulatory requirements and traditional high-priority concerns of the industry. The research agenda is developed through a process of consultation with subscribers and drinking water professionals. Under the umbrella of a Strategic Research Plan, the Research Advisory Council prioritizes the suggested projects based upon current and future needs, applicability, and past work; the recommendations are forwarded to the Board of Trustees for final selection. The foundation also sponsors research projects through the unsolicited proposal process; the Collaborative Research, Research Applications, and Tailored Collaboration programs; and various joint research efforts with organizations such as the U.S. Environmental Protection Agency, the U.S. Bureau of Reclamation, and the Association of California Water Agencies.

This publication is a result of one of these sponsored studies, and it is hoped that its findings will be applied in communities throughout the world. The following report serves not only as a means of communicating the results of the water industry's centralized research program but also as a tool to enlist the further support of the nonmember utilities and individuals.

Projects are managed closely from their inception to the final report by the foundation's staff and large cadre of volunteers who willingly contribute their time and expertise. The foundation serves a planning and management function and awards contracts to other institutions such as water utilities, universities, and engineering firms. The funding for this research effort comes primarily from the Subscription Program, through which water utilities subscribe to the research program and make an annual payment proportionate to the volume of water they deliver and consultants and manufacturers subscribe based on their annual billings. The program offers a cost-effective and fair method for funding research in the public interest.

A broad spectrum of water supply issues is addressed by the foundation's research agenda: resources, treatment and operations, distribution and storage, water quality and analysis, toxicology, economics, and management. The ultimate purpose of the coordinated effort is to assist water suppliers to provide the highest possible quality of water economically and reliably. The true benefits are realized when the results are implemented at the utility level. The foundation's trustees are pleased to offer this publication as a contribution toward that end.

Walter J. Bishop
Chair, Board of Trustees
Awwa Research Foundation

James F. Manwaring, P.E.
Executive Director
Awwa Research Foundation

ACKNOWLEDGMENTS

The authors of this report would like to thank the AwwaRF project manager, Alice Fulmer for all her support and encouragement during the project. The authors also wish to thank the Project Advisory Committee members, Sunil Kommineni (Malcolm Pirnie), Samuel Perry (Washington Department of Health), Eric Wert (Southern Nevada Water Authority), and Isai Urasa (Hampton University) for their valuable comments and suggestions.

A number of utilities and staff participated in the study by providing operational information and samples: James McDaniel, Pankaj Parekh, Gary Stolarik, Don Christie, Dat Quach (Los Angeles Department of Water and Power, California), James Taylor, Nick Herzog (City of Pomona, California), Michael Berlien, Chris Sanchez, Thomas Dennis (La Puente Valley County Water District, California), James Glancy, Eric Rogers (City of Lakewood, California), Ed Hansen, Lyle Ford (Magna Water Company, An Improvement District, Utah), Bill Gedney, Diane Gehling, John Hughes (Southern California Water Company, California), Jason Wen (City of Downey, California), Erin McCauley (California Water Service Company, California), Douglas Carrier (Indian Health Services, Arizona), Michelle De Haan (Damon S. Williams Associates, Arizona), Bob Oreskovich, Patrick Irwin (Dare County Water Department, North Carolina), Bill Ellis, Tony Perkins (Maine School Administrative District No.6, Maine), Roman Aguirre, Brian Primrose (US Filter, Colorado), and Paul Martyn (Los Angeles County Sanitation District, California). La Puente Valley County Water District and the City of Lakewood also provided the access to their sites and utilities and other in-kind services during the pilot studies.

Special thanks go to Rodney Mutter with EE&T who wrote the arsenic stabilization and minimization section of the proposal and a number of staff at Carollo Engineers for assistance with the project: Ryan Anderson and Hiriti Haileselassie for running the pilot tests and helping in finalizing this report; Jake Rennecker, Jason Davis, and Laura Baumberger with assistance during the bench testing for WWTP experiments; Jeff Berlin for assisting with residual regulations; Paul Bennett, Jason Berroteran, Stacy Fuller, Amanda Meier, Beth Morgan, Phat Nguyen, Jesse Thurston, and Barbara Tiedgen with preparing the document; Silvia Backlund, Donna Davis, Kelly Rellaford, Tony Ventura, and Lana Yao with graphics.

Finally, thanks to Gil Crozes, Robert Cushing with Carollo Engineers, and Bruce Macler with USEPA Region 9 for serving as technical advisors and reviewing the project deliverables.

EXECUTIVE SUMMARY

BACKGROUND

Recent regulatory changes have necessitated the development and evaluation of technologies used to treat arsenic and perchlorate in drinking water. Currently, the most widely used and accepted treatment technologies for the removal of arsenic, perchlorate, and nitrate involve the generation of a concentrated waste stream, which must be further treated or disposed. Depending on the local regulations, the disposal or discharge costs for these contaminants from the waste streams generated during drinking water treatment may be a significant fraction of the total project cost. Thus, it is important for utilities to consider all facets of the project to develop legitimate costs for dealing with these contaminants.

One of the objectives of this research is to identify different types of residuals generated from various treatment alternatives and the associated disposal or discharge approaches required to address arsenic, perchlorate, and nitrate-laden wastes. In addition, review of pertinent regulations and implementation issues associated with various treatment and disposal scenarios is critical to help utilities develop an overall framework in managing and addressing the necessary steps for implementing treatment systems. Arsenic treatment processes generate wastes and residuals that require a minimization approach. Since perchlorate and nitrate share similar chemical properties, these two contaminants are grouped together in developing residual management strategies. Thus, the project evaluated two distinct groups of waste minimization and treatment processes that effectively address different types of residuals.

This evaluation included strategies for the management of various types of arsenic wastes such as solid residuals from a conventional water treatment plant, backwash water from filtration processes, and liquid brine or concentrate from ion exchange or membrane processes. For perchlorate and nitrate treatment processes, types of wastes included brine from ion exchange, backwash and biomass waste from biological, concentrate from desalting membrane processes, and other solid wastes generated from phase separation processes (e.g., single use resin). This study identified a number of issues and steps that must be addressed in selecting an overall approach that can be best implemented for the treatment of these contaminants from source waters and the associated wastes. The final assessment included in the last chapter provides a basis for utilities to evaluate overall management strategies by identifying residuals and pertinent regulatory requirements.

PROJECT OBJECTIVES AND APPROACHES

As utilities are implementing arsenic, perchlorate, and nitrate treatment technologies to meet increasingly stringent regulations, implementing effective means for residuals management is becoming more important. The goals of this project were: (1) to develop and evaluate alternatives to minimize production of residuals and wastes generated during treatment of drinking waters containing arsenic, perchlorate, or nitrate, and (2) to treat residuals that are generated from the drinking water treatment processes. As new state and federal drinking water regulations are adopted to limit these contaminants in drinking water, critical evaluation of not only the primary treatment methods, but also the brine and waste management and disposal issues need to be thoroughly addressed to implement new treatment systems. Specifically, the objectives of this include the following:

- Identify different types of residuals generated from arsenic, perchlorate, and nitrate, treatment processes
- Identify approaches required to meet overall arsenic, perchlorate, and nitrate treatment goals in drinking water and residuals
- Develop and evaluate processes to minimize, manage, and treat concentrate waste streams
- Review pertinent regulations and compare costs and other issues associated with various treatment and disposal scenarios
- Summarize the results of this work in a format that will serve as a guidance tool for helping utilities develop effective residuals minimization and handling strategies.

In order to meet these objectives and provide useful guidelines for utilities coping with the issues mentioned, the following tasks were performed by the project team:

- Literature review
- Bench-scale minimization and stabilization of arsenic residuals and brine
- Bench-scale treatment of perchlorate and nitrate-laden wastes
- Pilot-scale minimization and recycle of arsenic backwash water
- Pilot-scale biological treatment of ion exchange brine
- Regulatory review
- Case studies and utility guidelines

The most promising technologies tested at the pilot scale were selected based on bench-scale testing results. The following is a list of goals for the pilot-scale tests:

- Confirm treatment efficacy or process optimization from bench-scale testing
- Identify potential issues associated with full-scale implementation
- Develop design and operating parameters for cost analysis
- Gather information to develop recommendations to the utilities

SUMMARY AND CONCLUSIONS

Arsenic Residual Treatment

- Bench and pilot-scale tests have shown that cartridge filters did not adsorb soluble arsenic, and that arsenic particulates, which were partially retained by cartridge filters, result from arsenic adsorption onto iron particles. Pilot tests showed that the filtration rate through cartridge filters would need to be lower than manufacturer-recommended (3 to 5 gpm/EQSL – equivalent single length) to minimize breakthrough.
- During the pilot study for arsenic treatment, high turbidity was observed in the spent backwash, possibly due to media carryover when the manufacturer-recommended

backwash rate of 10 gpm/ft^2 was used. This observation was substantiated by the elevated total iron concentration (up to 67 mg/L) in the backwash. The cartridge filters used did not perform as anticipated in removing arsenic from the recycle stream.

- Although not tested at the pilot-scale, coagulation-sedimentation performed well in the bench-scale tests by removing more particles than cartridge filters and by decreasing the total arsenic concentration in the spent backwash to levels lower than those measured in the groundwater. However, this approach would be more operator-intensive than cartridge or bag filtration, with regards to operation, maintenance, and waste handling.
- Based on the California Waste Extraction Test (WET) conditions, arsenic leaching from ferric chloride residuals (sludge) conditioned with 10 percent of lime or cement (weight basis) was reduced by 70 percent (from 20 to less than 6 mg/L). This reduction amount is higher than the 15 percent reduction, which is required in California to meet the regulatory limit of 5 mg/L of leached arsenic. The cost of landfilling hazardous materials will thus need to be compared by utilities to the cost of adding fixatives to meet non-hazardous criteria.
- In the calcium iron based (Ca-Fe) hydrogel sorbents, arsenic adsorption occurs throughout the media rather than only at the surface. This allows arsenic sorption densities to reach very high levels (nearly 1:1 on a molar basis with iron). Arsenic sorption was depressed in the presence of phosphate and sulfate at levels representative of ion exchange brines but arsenic adsorption was still observed even with these competing anions in large excess of arsenic. A limitation of the Ca-Fe sorbents, however, is their instability at neutral pH and in the presence of high bicarbonate concentrations. However, pH adjustment can be used to avoid the issue. The spent, dehydrated sorbents were stable under the conditions of the Toxicity Characteristics Leaching Procedure (TCLP) test and released between 0.3 and 7 percent of their arsenic content, depending on the dehydration conditions. Under the conditions of the WET test, an additional stabilization will be required.

Perchlorate and Nitrate Residual Treatment

- Under the conditions tested, successful biodegradation of nitrate and perchlorate was observed in brine streams from ion exchange treatment during both bench- and pilot-scale studies. High total dissolved solids (TDS) conditions reduced biomass production and decreased the reduction rates for both nitrate and perchlorate. Therefore, it may be necessary to maintain modest brine concentrations (2 to 3 percent) and moderately long residence times (12 to 16 hours) in suspended-growth, stirred-tank-reactor systems.
- Undiluted brine showed that typically nitrate could be biodegraded within 24 hours and nitrite within 48 hours. Complete perchlorate reduction was obtained in seven to ten days. The initial perchlorate reduction rates during the first two days of operation were in the range of 500 to 1,000 mg/L-day. If conditions are optimized to maintain this rate, complete perchlorate reduction in undiluted brine could be achieved in significantly less time.

- Based on the testing conducted on the ferric chloride spent ion exchange brine residuals, thermal technologies including spray roasting and spray drying can destroy perchlorate in the spent brine. In the spray-roasting environment (600 to 900°C), perchlorate is readily destroyed and residual hydrochloric acid is evaporated. Additional acid is also generated from pyrohydrolysis of ferric chloride. Thus, a net surplus of hydrochloric acid can be recovered from this process by hydrochloric acid absorbers to recycle regenerating acid solution. Spray drying is conducted at lower temperatures (550°C) and exhibits a significant temperature gradient with outlet temperature in the range of 100 to 200°C with similar performance.
- The bench-scale oxidation ditch simulation experiment demonstrated that bacteria indigenous to the Magna Water Company's wastewater treatment plant (WWTP) can reduce perchlorate using background organics as substrate. Perchlorate removal to below detection was achieved and sustained at empty bed contact times (EBCTs) as low as 30 minutes, a considerably shorter period than was required in the suspended mode (48 hours). Experiments at 5- and 10-minute EBCTs also demonstrated effective perchlorate removal, despite the small carbon bed volume used. However, hydraulic residence times of at least two days were required to achieve perchlorate degradations to below detection. Therefore, although the (an)oxidation ditch process did work, degradation kinetics must be enhanced before full-scale implementation of this or a similar process can be considered. This might be done through the addition of an exogenous electron donor such as methanol. However, given sufficient natural organic carbon present in the wastewater tested, methanol addition may not significantly improve perchlorate degradation kinetics.

RECOMMENDATIONS TO UTILITIES

Selection of the most appropriate technology is a challenge where a number of factors must be considered. A technology that provides good treatment performance in one location may not be suitable for other locations due to the limited discharge options, or distance to a disposal facility, etc. This project provides the treatment alternatives for residuals from arsenic, perchlorate, or nitrate treatment processes. Some of the results presented in the report are based on the technologies that are already available while new and innovative methods and concepts have also been demonstrated during the project. An important aspect of this project was to collect and disseminate information on residual treatment, which is sometimes overlooked in making a process selection.

In general, water quality for potential constituents needs to be thoroughly evaluated even if their levels may be lower than the drinking water guidelines. Many technologies concentrate contaminants in waste streams, and such concentration stop may result in levels in the residuals that exceed discharge or disposal regulations. As such, review of the technologies being considered not only on the drinking water treatment side, but also on the residual management side (disposal and discharge) is recommended to avoid any additional cost for residual management in the future. It is also important to recognize that each project is unique and that overall arsenic, perchlorate, and nitrate management requires a multifaceted approach that integrates technology, cost, regulation, and the public.

Specifically based on the results of the residuals management technologies evaluated in this project, the following recommendations are made to utilities considering implementation of residual treatment alternatives for arsenic, perchlorate, and nitrate.

- Pre-filtering raw water to minimize backwash frequency from single use media treatment for arsenic could extend the service run time and minimize backwash frequency to reduce waste discharge. The actual reduction in backwash frequency will depend on the quality and constituents in the raw water. It is recommended, however, that the cost of providing a pre-filtration step be compared with the cost of treating the spent backwash based on the results from this study. For treating backwash stream, the cartridge filters may be difficult to implement in backwash water treatment due to rapid headloss buildup caused by media carry over from some granular iron based systems. As such, further study of the coagulation-sedimentation process for backwash water from single use sorbent is recommended.
- Arsenic containing sludge was more effectively stabilized using lime based on the results from this study and previous research (MacPhee et al. 2001). It is recommended that lime be used for stabilizing arsenic containing residuals from coagulation sedimentation processes rather than cement (with or without superplasticizer). Cost of lime is lower than cement, and the water industry is already familiar with lime additions of up to 20 percent (w/w) and using filter presses for dewatering water treatment plant residuals for it to be a better alternative. Furthermore, elevated lime concentrations in the stabilized sludge may be valuable to the farming community if the stabilized residuals are land applied.
- The hydrogel sorbent used for solidifying arsenic brine waste showed sorption capacity as much as 1:1 arsenic to iron (molar basis) under the conditions tested. As indicated previously, in coagulation, much of the iron is in the bulk material of the iron oxyhydroxide floc and is unavailable for arsenic binding. With the Ca-Fe hydrogels, arsenic is adsorbed throughout the entire volume of the sorbents and a high proportion of the iron is available to bind arsenic. In contrast to coagulant sludges, the spent beads are easily dehydrated, producing a material that is amenable to handling and transport. The spent, dehydrated beads are compatible with the TCLP test but are not stable under conditions of the WET test. Thus, in California, some additional stabilization of the spent, dehydrated beads would be required prior to disposal. However, further testing of this sorbents, at pilot-scale, would be required to assess possible impacts of sorbent stability on performance over longer periods.
- High TDS conditions reduced biomass production and decreased the reduction rates for both perchlorate and nitrate. Therefore, it is necessary to maintain modest brine concentrations (2 to 3%) and moderately long residence times (12 to 16 hours) in suspended-growth, stirred-tank-reactor systems. Brine biodegradation conducted under this project did not address recovery and recycle of biomass to improve performance. However, other ongoing studies have shown that suspended growth processes, with mechanisms to retain biomass, significantly improved performance and enabled the biodegradation of brine at much higher concentrations. Biomass recovery would also reduce disposal surcharge costs for treated waste effluent. It is recommended that perchlorate biodegradation of this brine be evaluated using a biodegradation system that recovers and reuses biomass to minimize discharge volume. This can be accomplished by using technologies such as sequencing batch, which was demonstrated in this study, fixed film bioreactors, or adding a clarifier or a membrane filtration step. Biomass concentration and reuse should improve

perchlorate biodegradation performance in addition to minimizing effluent disposal and system operation costs.

- Another technology to greatly reduce perchlorate and nitrate waste is to use combination of selective resin and ferric chloride/hydrochloric acid based regeneration technology. Based on the testing conducted on ferric chloride residuals, two technologies were identified that could be implemented to achieve perchlorate destruction and partial recovery and reuse of components. These commercially available technologies are spray roasting and spray drying. It is recommended that alternative regeneration technologies using this regeneration technology for selective resin be considered and further evaluated. Such follow-up work will provide additional information on potential for decreased volume of generated wastes as well as recovery and reuse of regenerant solution components.
- Perchlorate and nitrate -laden concentrates from desalting membrane based plants combined with municipal wastewater may effectively be treated using a fixed-bed biological process as demonstrated in this project. The concept can also be applied as a regional solution where a number of such waste streams can be centrally treated. However, it is recommended that additional research be performed to resolve the remaining unknowns related to the design, start-up, and operation of such a process. Some of these unknowns include media selection, backwash protocol, impact of intermittent perchlorate exposure on degradation kinetics, and temperature effects. It is also recommended that studies be performed to investigate the efficacy of using a bioreactor to treat concentrate streams that have been blended with scalped municipal wastewater, an approach that might be more cost-effective than treating the entire wastewater flow.

FUTURE RESEARCH

A number of innovative technologies were investigated as part of the project. Although these technologies demonstrated successful treatment concepts for residuals containing arsenic, perchlorate, and nitrate, further research is recommended in the following areas.

Arsenic Residual Treatment

- Further tests investigating the treatment of backwash water from single use media using coagulation and settling at the pilot-scale are recommended. This approach may be more effective in treating arsenic in the backwash water for cases where backwash frequency is on the order of months.
- The biopolymer based Ca-Fe hydrogel sorbents performed well in solidifying the high concentration of arsenic in liquid brine. However, further testing of the sorbents, at pilot-scale, would be required to assess possible impacts of sorbent stability on performance over longer periods. The issue of stabilization for disposal in California requiring the WET test would also need to be addressed by optimizing the physical characteristics or by stabilizing dehydrated spent sorbents prior to disposal.

Perchlorate and Nitrate Residual Treatment

- Brine biodegradation conducted under this project did not address recovery and recycle of biomass to improve performance and efficiency of the system. However, other on-going studies have shown that suspended growth processes, with mechanisms to retain biomass, significantly improved performance and enabled the biodegradation of brine at much higher concentrations. Thus, continued research on biomass recovery would benefit the technology by optimizing the process.
- Pilot studies on thermal treatment of ferric chloride brine are recommended to validate perchlorate reduction performance, the effect of co-contaminants, and the form and quality of the iron product.
- Finally, although treatment of perchlorate at a WWTP was well demonstrated, further research is necessary to determine the optimal choice for the biogrowth support medium in a fixed-film reactor. Additional testing is required to determine whether or not perchlorate removal to below detection can be achieved and sustained at EBCTs shorter than 30 minutes with different support media.

CHAPTER 1
INTRODUCTION

BACKGROUND

Throughout the United States (U.S.), a number of utilities have been impacted by arsenic, perchlorate, and nitrate found in water supplies as more stringent regulations are being implemented. Arsenic occurs naturally in the environment, and it is also widely used in timber treatment, agricultural chemicals (pesticides), and manufacturing of gallium arsenic wafers, glass, and alloys. Perchlorate contamination is typically associated with facilities that once manufactured, handled, or stored ammonium perchlorate, a solid-rocket fuel, although other sources have been suggested for the cases where there is no identified source of contamination (Jackson et al. 2003). For nitrate, run-off from agricultural lands treated with nitrogen fertilizers has been recognized as the main source of occurrence.

Majority of the technologies for treating these three contaminants produce a concentrated waste stream that requires special handling and treatment prior to disposal. The disposal of these wastes can sometimes be very costly or not viable depending on the local discharge regulations. With the prevalence of these contaminants and newly proposed regulations, a research on innovative and cost effective approaches to deal with residuals discharge and disposal are warranted.

PROJECT OBJECTIVES AND APPROACHES

As more stringent state and federal drinking water regulations are adopted to limit these contaminants in drinking water, critical evaluation of not only the primary treatment methods, but also the residual management and disposal issues need to be carefully addressed to implement new treatment systems. Specifically, the objectives of this study were to:

- Identify different types of residuals generated from arsenic, perchlorate, and nitrate treatment processes.
- Identify approaches required to meet overall arsenic, perchlorate, and nitrate treatment goals in drinking water and residuals.
- Develop and evaluate processes to minimize, manage, and treat concentrate waste streams.
- Review pertinent regulations and compare costs and implementation issues associated with various treatment and disposal scenarios.
- Summarize the results of this work in a format that will serve as a guidance tool for helping utilities develop effective residuals minimization and handling strategies.

In Chapters 1 and 2, overview of regulations and current knowledge of drinking water treatment and residual treatment is presented. In subsequent chapters, specific residual treatment technologies that were investigated in this project are discussed in detail. Arsenic waste minimization, sludge stabilization, and brine solidification are discussed in Chapters 3, 4, and 5 respectively. Since perchlorate and nitrate share similar chemical properties, these two contaminants are grouped together in developing residual management strategies as shown in

Chapters 6, 7, and 8. The final assessment included in Chapter 9 provides a basis for utilities to evaluate overall management strategies by identifying residuals and pertinent regulatory requirements.

OVERVIEW OF DRINKING WATER REGULATIONS

Arsenic Drinking Water Regulation

In January 2001, USEPA promulgated a new standard for arsenic in drinking water that requires public water supplies to reduce the levels from 50 to 10 µg/L by 2006. In February 2002, the final rule became effective when USEPA adopted a 10 µg/L maximum contaminant level (MCL) for arsenic, which was previously at 50 µg/L. As a note, health-based federal maximum containment goal (MCLG) for arsenic is set at zero. In March 2003, California Office of Environmental Health Hazard Assessment (OEHHA) released a draft Public Health Goal (PHG) of 0.004 µg/L (4 ppt) for arsenic in drinking water based on the mortality of arsenic-induced lung and urinary bladder cancers observed in epidemiological studies of populations in Taiwan, Chile, and Argentina. This PHG and other factors such as residual treatment cost will be considered by the California Department of Health Services (CDHS) for developing a California arsenic standard. CDHS is expected to adopt a new state MCL for arsenic by 2004. CDHS must establish the MCL at a level as close as is "technically and economically feasible" to the PHG for the contaminant. In New Jersey, costs are not considered in setting the MCL for carcinogens. Thus, State of New Jersey has adopted a lower 5 µg/L MCL based on the analytical limitation, treatability, and risk factor. For most other states, compliance with the 10 µg/L MCL at a minimum will be required by 2006. A summary of pending and proposed arsenic regulations and health goals is shown in Table 1.1.

Perchlorate Drinking Water Regulation

Since perchlorate contamination has only been discovered recently, there are no state or federal standards for perchlorate in the U.S. USEPA has categorized perchlorate as needing further health effects and occurrence information before being considered for regulation under Contaminant Candidate List (CCL). Currently, USEPA is preparing to finalize a preliminary risk assessment and collecting occurrence data under the Unregulated Contaminant Monitoring Rule (UCMR). In March 2003, federal legislation (S.502) was introduced to facilitate setting a federal perchlorate MCL by 2004. However, under USEPA's current rule making schedule, this process was not finalized.

Table 1.1
Arsenic drinking water regulations

State	MCL	Note
USEPA and Most States	10 µg/L	USEPA MCLG is zero
California	Pending	Public Health Goal is set at 0.004 µg/L (4 ppt)
New Jersey	5 µg/L	Health Goal is set at 0.003 µg/L (3 ppt)

The National Center for Environmental Assessment has also completed a revised draft reference dose that is equivalent to 1 μg/L in drinking water. In 2003, the USEPA has requested a 15-month peer review panel at the National Academy of Sciences. The National Academy of Sciences will complete a review of the health implications of perchlorate ingestion, which is anticipated in 2004.

In January 1997, CDHS established a non-enforceable, advisory action level (AL) of 18 μg/L for perchlorate. CDHS used the upper value of 4 to 18 μg/L range that resulted from the "provisional" reference dose, which was in the federal regulatory framework. In January 2002, based on a draft risk assessment on perchlorate released by USEPA and the analytical detection limit, CDHS has reduced the AL for perchlorate from 18 μg/L to 4 μg/L. ALs are health-based advisory levels and not enforceable standards. When these limits are exceeded, water systems are required to notify local governing agencies and are recommended to issue a consumer notification. In addition, CDHS recommends that a source of drinking water be taken out of service if perchlorate concentration exceeds 10 times the current AL.

Due to the growing concern regarding the health effects of perchlorate in drinking water, California adopted legislation SB 1822 in September 2002 to establish the country's first drinking water standard for perchlorate by 2004. In December 2002, OEHHA released a draft PHG of 2 to 6 μg/L based on a human health study under the California Health and Safety Code §116275. The human data was evaluated by two approaches, a No Observed Adverse Effect Level (NOAEL) approach and a benchmark dose approach, thus proposing two PHGs at 2 and 6 μg/L (not a range but either 2 or 6 μg/L). As described, PHGs are concentrations of drinking water contaminants that pose no significant health risk if consumed for a lifetime, based on current risk assessment principles, practices, and methods. PHGs result from a "risk assessment," and do not consider economics or technical feasibility of removing the substance to these levels.

On March 11, 2004, the California OEHHA established a final perchlorate PHG of 6 μg/L. Due to the delay in establishing the perchlorate PHG, CDHS has not yet established a MCL for perchlorate. However, on March 11, 2004, CDHS announced that it will be using an AL of 6 μg/L for perchlorate, to be consistent with the recently established PHG, until the MCL is finalized.

Some states have adopted more stringent advisory levels for perchlorate. In 2002, Massachusetts Department of Environmental Protection (MDEP) issued a one-time state health advisory for perchlorate at 1 μg/L when very low levels of perchlorate were detected in several wells in the Bourne Water District (BWD) on Cape Cod. The wells are near the Massachusetts Military Reservation, where perchlorate has also been detected in large amounts and where an ongoing investigation and cleanup of groundwater contamination is in progress. MDEP recommends that if any system finds perchlorate at levels above 1 μg/L, it should provide a public notice to its "sensitive population subgroups" of customers, including pregnant women, infants, children up to the age of 12 and individuals with hyperthyroidism. In the notice, the affected customers are advised not to consume water that has perchlorate at levels exceeding 1 μg/L according to MDEP. BWD, however, made the decision to close any wells that contained perchlorate at 0.5 μg/L or higher. A summary of states that have adopted advisory guidance levels on health goals is provided in Table 1.2.

Table 1.2
Perchlorate drinking water regulations

State	Advisory Levels	Advisory Level Definition
Arizona	14 μg/L	Health Based Guidance Level based on child exposure (will be reviewed after USEPA's final Reference Dose (RfD))
California	6 μg/L	Action Level, Public Health Goal (18 μg/L initially then lowered to 4 μg/L before a new level was set)
Maryland	1 μg/L	Advisory Level (AL)
Massachusetts	1 μg/L	Precautionary Recommendation to a Local Water District for children and at-risk subpopulations
Nevada	18 μg/L	Public Notice Standard
New Mexico	1 μg/L	Recent Drinking Water Screening Level
New York	5 μg/L and 18 μg/L	2 Tiered Action Levels (planning and action)
Texas	4 μg/L	Action Level and Residential Cleanup Level (lowered from 22 μg/L)

Compiled based on Mayer 2003

Nitrate Drinking Water Regulation

The Federal MCLG for nitrate has been set at 10 mg/L (as N) or 45 mg/L (as NO_3^-) based on the health effects. The USEPA has set an enforceable MCL of 10 mg/L (as N), which is the lowest level to which water systems can reasonably be required to remove this contaminant should it occur in drinking water source. This federal regulation for nitrate became effective in 1992.

There is only a small margin of safety provided by the nitrate MCL. Also the level of nitrate in drinking water is subject to seasonal fluctuations beyond water systems control. Thus, utilities with nitrate levels exceeding 50% of the MCL (i.e., 5 mg/L as N) are required to include educational information regarding the risk posed by nitrate in their Consumer Confidence Reports (CCR).

In California, following OEHHA's review of the current literature and a reevaluation of USEPA's basis for calculating the MCLs for nitrate and nitrite, it was determined that there was no scientific basis to propose alternative PHGs in the State. Therefore, OEHHA has adopted the federal limits in establishing state's PHGs of 10 mg/L (as N) for nitrate, 1 mg/L (as N) for nitrite, and 10 mg/L (as N) for combined nitrate/nitrite in drinking water. The calculation of these PHGs was based on the protection of infants from the occurrence of methemoglobinemia, the principal toxic effect observed in humans exposed to nitrate or nitrite. In 1994, CDHS adopted the same limits for California's current drinking water standards (State MCLs) for nitrate (10 mg/L (as N)), for nitrite (1 mg/L (as N)), and for combined nitrate/nitrite (10 mg/L as N).

TREATMENT TECHNOLOGIES FOR ARSENIC, PERCHLORATE, AND NITRATE

As discussed previously, perchlorate, and nitrate share similar chemical characteristics. Thus, a treatment process that can be implemented for perchlorate can also be used for nitrate in general. A number of removal technologies are available to treat either perchlorate or nitrate, and treatment technologies can be generally classified as phase transfer processes or destruction processes. Technologies such as ion exchange, granular activated carbon (GAC), high-pressure membrane treatment, and electrodialysis reversal (EDR) transfer perchlorate and nitrate from one phase to another and generate a waste stream that must be managed. On the other hand, biological processes degrade perchlorate and nitrate to chloride and nitrogen, respectively and do not produce any contaminant-laden waste streams. Some of these technologies described can also be implemented for arsenic. However, the approach for arsenic will be different as all options available will generate arsenic-laden waste.

A summary of these treatment technologies that are used or have been investigated for drinking water applications is presented in Table 1.3. It should be noted that the description given in the table are for general comparison only, and applicability of each technology will be case specific.

Table 1.3
Comparison of treatment processes for arsenic, perchlorate, and nitrate

Process	Applicability	Advantages	Disadvantages
Media adsorption (single use)	Arsenic	- High selectivity for arsenic. - Generally no chemical is required (if pH adjustment is not needed).	- Spent media must be replaced on a regular basis. - Competing anions (vanadium, etc.) may reduce treatment efficiency.
Media adsorption (regenerable – on site or off site)	Arsenic	- Media can be re-used (for a limited time). - Simple operation if regeneration is done off-site.	- Chemical regeneration is required. - pH adjustment is required for the waste regenerant (for AA). - Some regenerable media are still under filed testing (iron-incorporated resin, TiO2 media)
Coagulation/ filtration (media, membrane, etc.)	Arsenic	- Effective, proven technology. - Low costs (for media filtration).	- Efficiency significantly reduced at high or low pH. - Disposal of arsenic laden sludge may be a concern.

(continued)

Table 1.3 (Continued)

Process	Applicability	Advantages	Disadvantages
Conventional treatment	Arsenic	- Effective, proven technology applicable for a large scale use.	- Efficiency depends on pH. - Produces a large quantity of arsenic laden sludge. - Dewatering of sludge is required before disposal.
Ion exchange (regenerable)	Arsenic Perchlorate Nitrate	- Proven technology. - Can remove arsenic, perchlorate, nitrate, and other anions at the same time. - Resins are re-used after regeneration. - Potentially high rate of treatment.	- Some resins may produce precursors to form NDMA in finished water. - Efficiency depends on raw water quality. - Generates brine with high TDS, arsenic, perchlorate, and nitrate.
Membrane (RO/NF)	Arsenic Perchlorate Nitrate	- Can achieve rejection of multiple contaminants. - Proven technology for drinking water. - Added benefits such as hardness reduction.	- High capital and O&M costs. - Generates a large quantity of concentrate waste containing arsenic, perchlorate, and nitrate. - High TDS and silica reduce efficiency of removal. - Long-term removal of arsenic using RO/NF has not been proven, yet. Anecdotal studies have shown inconsistent arsenic removal using NF/RO.
Membrane (EDR)	Arsenic Perchlorate Nitrate	- Can achieve rejection of multiple contaminants. - Proven technology for drinking water. - Can be applied to water with high TDS and silica.	- High capital and O&M costs. - Generates a large quantity of concentrate waste containing arsenic, perchlorate, and nitrate.

(continued)

Table 1.3 (Continued)

Process	Applicability	Advantages	Disadvantages
Biological	Perchlorate Nitrate	- Complete destruction of perchlorate and nitrate. - Can achieve multiple treatment objectives including perchlorate, nitrate, and some organics. - Indigenous micro-organisms can be used. - Relatively low cost (O&M) for perchlorate removal.	- Public acceptance. - No current full-scale applications for direct drinking water treatment (in the U.S.). - Requires a post-treatment train for potable water applications. - Requires NSF certified electron donor. - Acclimation period is required to develop biological culture within the bed.
Ion exchange (single use)	Perchlorate	- Well-demonstrated technology. - Does not produce perchlorate-laden liquid brine. - Can be easily implemented.	- Resins must be replaced on a regular basis (every few months). - High O&M costs associated with resin replacement. - Some resins may produce precursors to form NDMA in finished water. - Presence of uranium or hazardous constituents may generate hazardous or low level radioactive waste.
GAC/ modified GAC	Perchlorate	- No perchlorate-laden liquid waste is generated. - Existing facility (GAC facility) may be used.	- Limited perchlorate adsorption capacity without modification of GAC. - GAC must be replaced once adsorption capacity is saturated. - Monomer or polymer used to modify GAC may leach and associated DBP formation is not well known. - CDHS has not accepted modified GAC (CDHS has conditionally accepted GAC).

IDENTIFICATION OF TYPES OF RESIDUALS

The residuals produced from the treatment processes listed in Table 1.1 and the potential disposal methods for the residual waste streams are identified in Tables 1.4 and 1.5. Table 1.4 lists technologies and residuals associated with arsenic treatment, and Table 1.5 summarizes similar information for perchlorate and nitrate. However, actual waste handling related to discharge or disposal of the wastes is subject to local, state, and federal regulations depending on the ultimate location of the discharge/disposal point. More discussion on the regulations governing arsenic, perchlorate, and nitrate- laden wastes is found in the following section.

Table 1.4
Identification of waste from arsenic treatment processes

Treatment technology	Form of residual	Type of residual	Arsenic in the waste stream?	Potential disposal methods
Ion exchange (regenerable)	Liquid	Spent brine	Yes	Treat and discharge to brine line or discharge to holding tank and haul
		Rinse/backwash waste	No	Discharge to sewer or same options as spent brine if mixed with brine
	Solid	Spent resin	No	Disposal to landfill
Media adsorption (single use)	Liquid	Backwash	Yes	Discharge to sewer if arsenic level is low
	Solid	Spent media	Yes	Disposal to landfill
Media adsorption (regenerable)	Liquid	Backwash	Yes	Treat (if necessary) and discharge to sewer
		Spent regeneration solution	Yes	pH adjustment and discharge to sewer
	Solid	Spent media	No	Disposal to landfill
Coagulation filtration (media or microfilter)	Liquid	Backwash	Yes	Discharge to sewer or evaporation pond
	Solid	Settled sludge (if separated from backwash)	Yes	Disposal to landfill

(continued)

Table 1.4 (Continued)

Treatment technology	Form of residual	Type of residual	Arsenic in the waste stream?	Possible disposal methods
Conventional treatment	Liquid	Backwash	No	Discharge to sewer
	Solid	Settled sludge	Yes	Disposal to landfill
Membrane (RO, NF, and EDR)	Liquid	Concentrate	Yes	Discharge to sewer, evaporation pond, or limited groundwater injection
		Membrane cleaning waste	No	pH adjustment and discharge to sewer

Modified from EPA 2000

Table 1.5
Identification of waste from perchlorate and nitrate treatment processes

Treatment technology	Form of residual	Type of residual	Perchlorate or nitrate in the waste stream?	Possible disposal methods
Ion exchange (single use)*	Liquid	Resin transfer water, initial rinse water	No	Discharge to sewer
	Solid	Spent resin	Yes	Incineration or disposal to landfill
Ion exchange (regenerable)	Liquid	Spent brine	Yes	Discharge to brine line or discharge to holding tank and haul
		Rinse/backwash waste	No	Discharge to sewer or same options as spent brine if mixed with brine
	Solid	Spent resin	No	Disposal to landfill
Biological (fixed and fluidized bed)	Liquid	Backwash or biomass waste	No	Discharge to sewer

(continued)

Table 1.5 (Continued)

Treatment technology	Form of residual	Type of residual	Perchlorate or nitrate in the waste stream?	Possible disposal methods
Biological (membrane biofilm)	Liquid	Biomass waste	No	Discharge to sewer
		Membrane cleaning waste	No	Discharge to sewer
Membrane (RO, NF, and EDR)	Liquid	Concentrate	Yes	Discharge to sewer, evaporation pond, or limited groundwater injection
		Membrane cleaning waste	No	pH adjustment and discharge to sewer
GAC/modified GAC*	Liquid	Backwash	No	Discharge to sewer
	Solid	Spent carbon	Yes	Return to supplier for regeneration

* Perchlorate treatment only where noted

OVERVIEW OF DISCHARGE AND DISPOSAL REGULATIONS

Arsenic Discharge and Disposal Regulations

A summary of arsenic residual management methods is provided in Table 1.5. These arsenic management methods are classified into two different groups based on whether they are used to manage liquid residuals or solid and sludge residuals. Some of the liquid residual disposal methods are direct discharge to water body, indirect discharge via a sanitary sewer system, underground injection, management in lagoons, and possibly land disposal or land application. Different federal regulations apply to each liquid residual disposal method. Solid and sludge residuals are usually disposed through landfilling, land application, storage in lagoons, and recovery, or disposal of spent media. Accordingly, different federal regulations apply to solid waste landfills, hazardous waste landfills, lagoons, reuse of hazardous waste, reuse of solid waste, and off-site disposal. Note that typical disposal methods for liquid residuals are direct discharge to a water body and indirect discharge via a sanitary sewer system, whereas the typical disposal methods for solid and sludge residual are landfilling, in municipal or industrial landfills, and land application.

In general, the key factors that drive the applicability of federal regulations on arsenic removal residuals are the physical form of the residual (liquid or solid and sludge residual), the method(s) used to manage the residual, and whether or not a residual constitutes a hazardous waste or a non-hazardous waste as defined under the Resource Conservation and Recovery Act (RCRA). Figure 1.1 depicts how these three variables can be used to determine which federal regulations apply to the management of arsenic residuals.

In addition to the federal regulatory requirements, arsenic-laden residuals are also regulated by the state regulatory programs. Many of the state regulatory programs are generally consistent with the federal minimum requirements. Nevertheless, state programs differ from federal requirements and each other in several important aspects. These differences include the actual surface water quality standards applicable to control the amount of arsenic in direct discharge of liquid effluent, the local limits that specify how much arsenic may be discharged to a sanitary sewer system, the regulation of solid waste landfills, the protection of ground water resources, and the regulation of land application activities. Thus, in addition to the federal regulations, it is important to have a clear understanding of the state regulatory programs that are applicable to the management of arsenic removal residuals.

As mentioned earlier, typical disposal methods for liquid residuals are direct discharge to a water body and indirect discharge via a sanitary sewer system. In the management of liquid residual through direct discharge to surface waters, the most relevant state standards are state water quality standards for selected states, which are presented together with the recommended national numeric water quality criteria for arsenic through National Pollutant Discharge Elimination System (NPDES) permit. In the case of California, the regional basin plans also impose numeric surface water quality standards. Most of the surface water quality standards

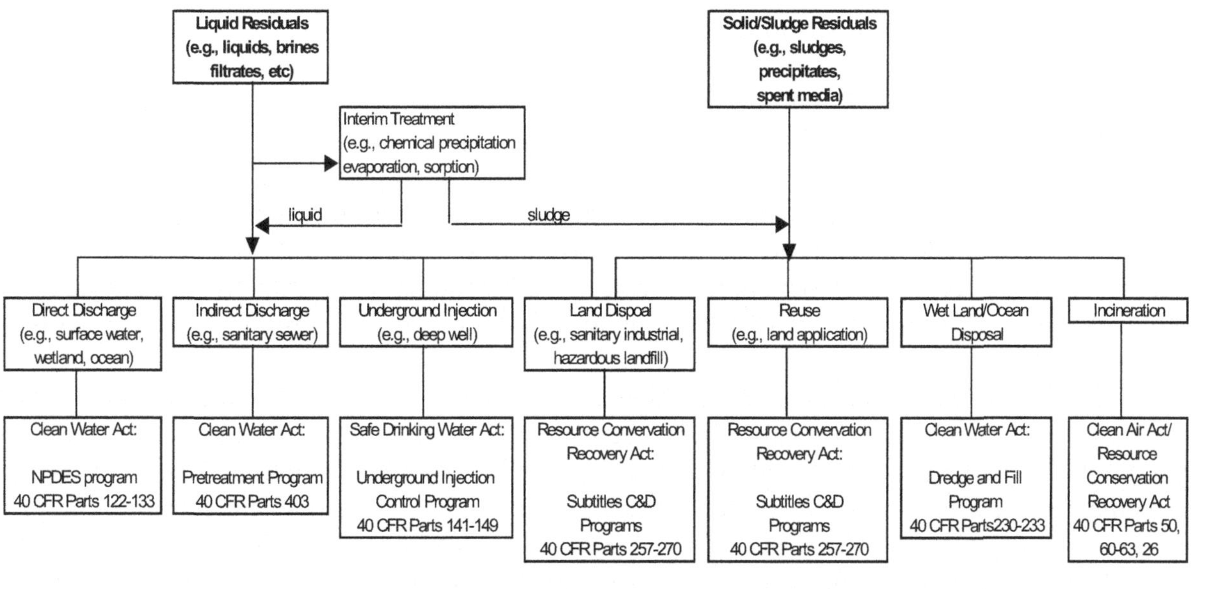

Figure 1.1 Federal regulations governing the disposal of arsenic residuals (USEPA 2000)

imposed under the statewide plans are replaced by the California Toxics Rule (CTR). In the case of indirect discharge to a sanitary sewer system, the key regulatory standard is the local arsenic discharge limit to sanitary sewer system. In Arizona, the local arsenic discharge limit to Publicly Owned Treatment Works (POTW) is 0.1 mg/L. In the State of New Mexico, the local discharge limits range from 0.051 mg/L in Albuquerque to 2.74 mg/L in Santa Fe. In the case of solid/sludge residual, the typical disposal methods are solid waste landfills, hazardous waste landfills, and land application (reuse). Solid waste landfills must comply with arsenic groundwater monitoring requirements and other design and operating provisions. These requirements are implemented at the state level and, as noted, have been adopted by at least 40 states. Depending on the leaching test level, solid waste is disposed of at various class landfills (i.e., Class I, II, or III). Most of the federal regulation for solid and sludge arsenic treatment residuals occur under RCRA subtitle D (non-hazardous waste). In cases where arsenic treatment residuals constitute hazardous waste, the residuals must be managed according to RCRA Subtitle C requirements (hazardous waste). For hazardous waste landfills, Arizona, California, Maine, Nebraska, New Mexico, Nevada, and Pennsylvania are some of the states that adopt the federal arsenic standard for the purpose of the toxicity characteristics (i.e., 5.0 mg/L). If the extract from a representative sample of the waste (based on use of the TCLP) exceeds this threshold, the waste is regulated as a hazardous waste. California also has developed an additional leaching test known as the waste extraction test (WET). Based on the WET results, California defines those wastes that have an arsenic soluble threshold limit concentration (STLC) of 5.0 mg/L and those wastes that have a total threshold limit concentration (TTLC) of 500 mg/kg, as a hazardous waste. The land application of drinking water treatment plant (DWTP) sludge appears to be regulated primarily on a case-by-case basis under the respective state regulations. More detailed information can be found in the USEPA document (USEPA 2000) on this subject and general discharge and disposal of arsenic-laden residuals.

Perchlorate Discharge and Disposal Regulations

Currently, there is limited information available on perchlorate discharge regulation, in part because there are no established state or federal drinking water regulations for perchlorate. In California where a number of treatment plants produce perchlorate-laden ion exchange brine, different agencies are involved with regards to regulating disposal of the perchlorate waste. As part of the superfund clean up activities, several utilities in San Gabriel area in Los Angeles County are required to treat perchlorate brine by 2006 under USEPA's Record of Decision (RoD) for that Superfund program. This regulation and timeline is currently limited to utilities involved in that Superfund program.

In 2003, Los Angeles County Regional Water Quality Control Board adopted 4 μg/L perchlorate discharge limit to the NPDES permit. This limit was set to the action level for perchlorate at that time. Shortly after that, Los Angeles County Sanitation District (LACSD) implemented the same 4 μg/L maximum discharge limit to their system where ocean discharge is not available. For areas where the ocean discharge is available, LACSD does not limit the amount of perchlorate that is discharged into their line as long as the contaminant does not originate from a Superfund site.

On the other hand, Orange County Sanitation District (OCSD) currently regulates perchlorate discharge given that it is an emerging contaminant and environmental effects are unknown. There is no set limit on the amount of perchlorate that can be discharged to OCSD. Permits are granted for discharge on a case-by-case basis. Before a permit can be granted for the discharge of perchlorate into the system, a letter pertaining to the discharge characteristics needs to be submitted to OCSD for their review of the project. Important parameters to be considered by OCSD include duration of project, minimum and peak flows, minimum and peak perchlorate levels in waste, characterization of source waters. These criteria are reviewed to set the discharge limit.

In Kansas, the Kansas Department of Health and Environment allows a manufacturing facility (Kerr McGee) to discharge up to 100 μg/L of perchlorate to a WWTP. It is not known if other states impose any further restrictions on perchlorate discharge.

Among the residuals identified for perchlorate in Table 1.5, the single use ion exchange and GAC/modified GAC processes are the only technologies that generate solid wastes that need to be disposed of. Currently, the modified GAC is still under development, and it is not clear how the reactivation or disposal of spent carbon will take place. The spent resin generated from the single use ion exchange process is typically disposed by incineration. Although no perchlorate disposal regulation is in place, the disposal of perchlorate by destruction (incineration) of the spent resin ensures the utilities from potential future liabilities associated with perchlorate disposal. This service and the cost are generally included in the resin change out service provided by the equipment manufacturers.

Depending on the operation and the raw water quality, prolonged use of the resin may accumulate other constituents such as other anions (e.g., arsenic) and radioactive compounds. If this is the case, then other regulations pertaining to hazardous or low level radioactive status may apply. To prevent such case, the water quality must be thoroughly characterized and the design and operation of single use system must be chosen such that accumulation of other constituents in the spent resin can be avoided. The spent resin is profiled (profiling is the testing of a waste to determine that the waste contains what is claimed by the generator) for disposal. The spent resin is then sent to a certified laboratory for WET and TLCP analysis. Analytical results determine hazardous or non-hazardous status of the spent resin. Since perchlorate saturated resin alone is not considered a hazardous waste, the analysis is performed to determine if other toxic species may be present. If hazardous waste is present due to the perchlorate loading, the spent resin is disposed by RCRA incineration. Additional measures also need to be taken if unidentified contaminants make the spent resin hazardous. The spent resin is then disposed after being consolidated with other fuel blending wastes at which time a certificate of destruction (CoD) is issued to a utility.

Nitrate Discharge and Disposal Regulations

There are no established regulations for discharging nitrate wastes, although there may be some local restrictions based on the regional basin management goal. USEPA is progressing with "Ecoregional Nutrient Criteria" for surface water quality. These criteria are not necessarily applicable for regulating nitrate discharge, but they may provide insight for potential future requirements, based on "ecoregion" location for discharging of nitrate waste. Where ocean discharge is available, a different set of regulations may apply. In California, ocean discharge is regulated by the California Ocean Plan developed by the State Water Resources Control Board (SWRCB 2001). Nitrate is not listed specifically by the Ocean Plan, but a narrative requirement states that "Nutrient materials shall not cause objectionable aquatic growths or degrade indigenous biota."

In California, OCSD with ocean discharge does not currently limit the nitrate into the brine line known as Santa Ana Regional Interceptor (SARI) line. However, OCSD is expected to add constituents and adjust local limits as it transitions toward implementation of the Groundwater Replenishment System (GWRS). These additions and limits are based upon the CDHS Title 22 requirements since the treated GWRS water will be percolated into the Orange County groundwater basin. Further, Santa Ana Watershed Project Authority (SAWPA) is beginning an evaluation of the SARI System Enhancements Program that, if implemented, would require treatment of nitrate prior to discharge into the SARI, which would then discharge directly into the OCSD outfall, bypassing OCSD's WWTP (Plant No. 2). In addition, if Plant No. 2 becomes a reclamation plant, limits could be imposed for nitrate by CDHS. Similar local and state regulations may be imposed elsewhere where such approaches are implemented.

The treatment processes summarized earlier for nitrate would not produce any solid waste-containing nitrate. Thus, nitrate disposal regulation for solid waste would not be applicable for this case.

CHAPTER 2
TREATMENT AND MINIMIZATION TECHNOLOGIES FOR RESIDUALS

OVERVIEW OF RESIDUAL TREATMENT

One of the arsenic treatment processes often utilized by large utilities is chemical precipitation. This process produces a large quantity of solid residuals that must be disposed. The majority of utilities affected by the new arsenic rule, however, have small scale applications that use phase transfer technologies including adsorption or membrane removal processes, primarily activated alumina (AA), iron based sorbents, ion exchange (IX), reverse osmosis (RO), nanofiltration (NF), or electrodialysis reversal (EDR). Each of these processes generates liquid residuals such as arsenic-laden liquid brine, backwash water, and/or concentrate that may require treatment before disposal. Typical approaches for dealing with liquid arsenic wastes are chemical precipitation (Clifford and Lin 1995) or sorption using media filters, such as AA or iron modified sorbents (MacPhee et al. 2001). Some research has been done on controlling arsenic speciation and solubility in sewage sludge by promoting biomethylation and production of organic arsenic compounds with a much lower toxicity than their inorganic counterparts (Carbonell-Barrachina et al. 2000). The World Health Organization recognized this process as one of the options to deal with arsenic-containing waste (WHO 2001). However, arsenic biomethylation is difficult to sustain, and since the current regulation does not factor in toxicity, there is little advantage in using this process to meet current regulations. Table 2.1 describes a number of technologies investigated for stabilizing and minimizing various types of arsenic-containing wastes.

Table 2.1
Arsenic stabilization/minimization technologies

Technology	Scale	Waste stream	Description	Reference
Lime addition	Pilot	Sludge	Blend lime with residuals to prevent arsenic leaching.	MacPhee, Cornwell, and Mutter 2003
Fixatives	Pilot	Sludge	Fix/stabilize sludge with plasticizers to prevent arsenic leaching.	Gallup 2002
Bio-methylation	Full	Sludge	Conversion of inorganic arsenic species to less toxic organic species under anoxic conditions in treatment plant sludge.	Carbonell-Barrachina et. al. 2000
Sorption media	Pilot	Liquid	Use of sorption medias to treat liquid brine from ion exchange processes.	MacPhee, Charles, and Cornwell 2001

(continued)

Table 2.1 (Continued)

Technology	Scale	Waste stream	Description	Reference
Chemical precipitation	Pilot	Liquid	Add ferric chloride to ion exchange brine to stabilize liquid waste containing arsenic.	Clifford and Lin 1995
Hydrogel sorbent	Bench	Liquid	Selective sorption or arsenic and solidification of hydrogel upon saturation to reduce volume and weight.	Min and Hering 1998a, 1998b
Backwash recycle	Pilot	Liquid	Recycle of backwash water to minimize arsenic containing residual.	Narasimhan, Padilla, and Devkota 2002

Several studies have demonstrated perchlorate and nitrate destruction under various experimental conditions as shown in Table 2.2 and Table 2.3. Unlike arsenic, which must be separated from the bulk solution before treatment, perchlorate and nitrate can be degraded by biological or other catalytic processes. These processes can thus be used directly to treat perchlorate and nitrate in water without generating a concentrated waste stream. Alternatively, they can be used in conjunction with a separation processes to treat the waste that is generated during drinking water treatment. Since a number of research projects funded by AwwaRF have addressed direct perchlorate destruction technologies, the focus of this proposed study is on treating brine streams generated during separation processes, namely ion exchange.

Table 2.2
Biological perchlorate and nitrate destruction technologies

Technology	Scale	Application (direct or residuals)	Influent perchlorate conc.	Effluent perchlorate conc.	Reference
Biologically active carbon (BAC) filtration*	Bench	Direct	10-300 μg/L	< 2 μg/L	Brown, Snoeyink, and Kirisits 2002
Ozone and BAC filtration*	Bench & Pilot	Direct	10-100 μg/L	< 4 μg/L	Liang, Min, and Snoeyink 2000
Hollow fiber membrane biofilm reactor*	Bench & Pilot	Direct	6-100 μg/L	< 2 μg/L	Nerenberg and Rittmann 2001

(continued)

Table 2.2 (Continued)

Technology	Scale	Application (direct or residuals)	Influent perchlorate conc.	Effluent perchlorate conc.	Reference
Packed-bed bioreactor	Bench	Direct	740 μg/L	460 μg/L	Miller and Logan 2000
Culture tube bioreactors	Bench	Residuals	100 mg/L (1.5% NaCl conc.)	90 mg/L	Gringas and Batista 2002
Batch bioreactors	Bench	Residuals	592 mg/L (3% NaCl conc.)	45 mg/L	Logan, Wu, and Unz 2001
Suspeneded bioreactor	Full	Waste	variable	< 500 μg/L	Coppola et al. 1998
Fluidized bed reactor*	Full	Direct	2.7 mg/L	< 4 μg/L	Guarini 2002
Biological – Baldwin park OU*	Pilot	Direct	50-100 μg/L	< 4 μg/L	Catts 1998
Celite-packed columns*	Bench	Direct	738 μg/L	< 4 μg/L	Giblin et al. 2000

*Both perchlorate and nitrate removal

Table 2.3
Non-biological perchlorate and nitrate destruction technologies

Technology	Scale	Application (direct or residuals)	Influent perchlorate conc.	Effluent perchlorate conc.	Reference
UV w/catalyst	Bench	Direct	1,000 μg/L	200 μg/L	Kim and Gurol 1999
Pulsed-UV w/ catalyst	Pilot	Direct	variable	variable	Mofidi et al. 2002
Oxorhenium(v) coordination complex catalysis with sulfides	Bench	Direct	9.8 g/L	not given	Abu-Omar et al. 2000
Titanous ion catalysis in ethanolic solutions	Bench	Direct	4.0 g/L	not given	Early et al. 1999

(continued)

Table 2.3 (Continued)

Technology	Scale	Application (direct or residuals)	Influent perchlorate conc.	Effluent perchlorate conc.	Reference
Catalytic perchlorate and nitrate destruction module*	Pilot & Full	Residuals	variable	< 4 μg/L	Venkatesh et al. 2000
Thermal destruction in GAC	Bench & Pilot	Direct	variable	< 4 μg/L	Cannon and Chongzheng 2000
Thermal perchlorate decomposition system (600°C)	Bench	Residuals	50 mg/L (7% NaCl conc.)	< 500 μg/L	Li and Coppola 2000
Hydrothermal perchlorate decomposition system (380°C)	Bench	Residuals	50 mg/L (7% $FeCl_3$ conc.)	< 500 μg/L	USEPA 1999
γ-ray	Bench	Direct	NA	NA	Hart et al. 1980
Electron beam	Pilot	Direct	variable	variable	Deshmukh et al. 2002
Electro-catalytic conversion of nitrate to nitrogen gas #	Pilot	Residual	12 mg/L	2.5 mg/L	Ionex 2000

*Both perchlorate and nitrate removal
#Nitrate removal only

ARSENIC BACKWASH WATER MINIMIZATION AND RECYCLE

With regards to arsenic removal from drinking water, a number of papers have been published detailing the treatment effectiveness of both conventional treatment (coagulation and softening) and phase transfer processes (ion exchange, RO, and AA). Both aluminum (Al) and iron salts are effective for removing arsenic, but factors such as pH, arsenic oxidation state, and competing ions affect the removal efficiency.

Apart from the recent projects conducted by EE&T for USEPA (MacPhee et al. 2001) and AwwaRF (MacPhee et al. 2003), a review of the literature yielded relatively few works that specifically address characteristics of residuals containing arsenic and removal of arsenic from those residuals. Ela et al. (2003) and Ghosh et al. (2002) indicated that actual landfill conditions (e.g., alkaline pH, anaerobic microbial activity, high phosphate, concentrated organic, etc.) are not reflected in standard leaching tests such as TCLP and WET. The research showed that using actual and simulated landfill leachates instead of the extraction fluids used in the TCLP and WET procedures resulted in higher arsenic leaching from AA and GFH adsorbents. For both sorbents the mobilization of arsenic increased with pH, with critical desorption occurring in pH between 8 and 10 for AA, and pH between 6.5 and 9 for GFH. Anaerobic bacteria also reduced arsenic (V) to arsenic (III) within seven days. The authors recommended that future residuals assessments be developed by integrating such impacts as pH, microbial processes, and metal speciation under relevant redox conditions.

Hathaway and Rubel (1987) and Clifford and Lin (1986) both reported results of Extraction Procedure (EP) toxicity tests performed on residuals containing arsenic. Recent USEPA publications (Wang et al. 2000, Fields et al. 2000) present TCLP results for residuals collected at arsenic removal treatment facilities. Other publications, including Chen et al. (2002), Clifford et al. (1998), and Chwirka (1999), address levels of arsenic in residuals. However, the residuals characterization performed in these works did not address the periodic backwashes of AA and GFH sorbents for removal of headloss-causing solids before the media is actually exhausted.

MacPhee et al. (2001) demonstrated adsorption tests for different types of media and resins to remove arsenic from liquid residuals but did not assess ultimate capacity. AA and iron-based media, such as GFH, provided comparable arsenic removals of close to 100% with an empty bed contact time (EBCT) of 3 minutes for most of the membrane concentrates and the settled iron removal facility residuals. Removal of suspended solids was key to the success of adsorption for spent filter backwash water and clarifier flush residuals. Alkalinity significantly interferes with adsorption of arsenic. Based on this work, use of adsorption media for treatment of arsenic-laden water plant residuals merits further exploration. TCLP arsenic levels in all residuals generated in this work and in full-scale solid media samples were less than 0.5 mg/L, which is far below the regulatory limit of 5 mg/L.

In MacPhee et al. (2003), spent filter backwash water (SFBW) samples from six different utilities were analyzed for total arsenic, total iron, total aluminum, TSS, pH, and turbidity. Results from these tests are presented in Table 2.4. The data show that three of the six samples had arsenic concentrations greater than 50 μg/L. Without solids removal, these samples would exceed the NPDES limit for direct discharge. Depending on the local POTW and the Technically Based Local Limit (TBLL) established for arsenic, these residuals may or may not be acceptable for indirect discharge.

Table 2.4
Results of SFBW characterization analysis

Utility name	Type	Turbidity (NTU)	pH	Total As (μg/L)	Total Fe (mg/L)	Total Al (mg/L)	Ca (mg/L)	Mg (mg/L)	TSS (mg/L)
Louisiana Water Co. (New Iberia), La.	Lime	1,150	8.6	43	18.6	1.9	267	18	2,364
City of Great Falls, Mont.	Alum	7.8	7.5	32	0.1	4.8	-	-	269
City of Helena, Mont.	Alum	265	7.1	163	9.2	21.9	-	-	515

(continued)

Table 2.4 (Continued)

Utility name	Type	Turbidity (NTU)	pH	Total As (μg/L)	Total Fe (mg/L)	Total Al (mg/L)	Ca (mg/L)	Mg (mg/L)	TSS (mg/L)
Billings Dept. of Public Utilities (Billings), Mont.	Ferric	28	7.7	63	34.6	1.2	-	-	119
Lockwood Water Users Assoc. (Lockwood), Mont.	Ferric	12	7.3	14	8.7	0.4	-	-	28
Los Angeles Dept. Of Water & Power (LADWP), Calif.	Ferric	208	7.2	185	66.3	5.6	-	-	

Data from MacPhee et al. 2003

The data shown in Table 2.5 indicate that effective arsenic removal was achieved by treatment of backwash water with either $FeCl_3$ or polymer, followed by gravity settling. The removal of particulate associated turbidity also removed arsenic from the backwash water.

MacPhee et al. (2003) recommended that utilities in the process of selecting an arsenic removal treatment technology evaluate the types of residuals that would be generated, their expected arsenic concentrations, and pre-treatment strategies that would be required prior to final disposal. Arsenic can be removed from a liquid residual stream using coagulation and clarification, or adsorption prior to disposal.

Table 2.5
Comparison of best SFBW treatments using polymer and $FeCl_3$ (MacPhee et al. 2003)

	No chemical treatment	Polymer		Ferric chloride	
Utility name	As remaining (μg/L)	Dose (mg/L)	As remaining (μg/L)	Dose (mg/L)	As remaining (μg/L)
New Iberia, La.	30	1	7	120	5
Great Falls, Mont.	29	4	6	120	<2
Helena, Mont.	159	4	<2	20	3
Billings, Mont.	24	3	<2	40	<2
Lockwood, Mont.	10	3	4	40	<2
LADWP, Calif.	50	0.5	<2	20	7

Data from MacPhee et al. 2003

ARSENIC SLUDGE STABILIZATION

In MacPhee et al. (2003), a project was structured to determine arsenic release from residuals due to chemical conditioning for dewatering, thickening, short- and long-term storage, as well as to determine innovative methods for preventing arsenic mobility from residuals into the environment. Characterization and laboratory treatability work were used to develop a database comparing paired TCLP and California WET results for the same residuals. These residuals were collected from full-scale coagulation, lime softening, and iron/manganese removal processes.

Each sample was divided into three portions to generate the different residuals streams required for analysis. Settled residuals collected from sedimentation basins (or from a lagoon in the case of Los Angeles Department of Water and Power's (LADWP) direct filtration facility) were completely mixed as delivered, and were referred to as untreated residuals. Secondary residuals were created from the separation of the liquid and solid portions. These residuals were gravity settled to separate the solids from the liquid portion and then the solids were either air- or oven-dried prior to analysis. The results from the chemical characterization analyses are summarized in Table 2.6.

The data also indicated significant differences between the TCLP and WET arsenic concentrations measured from the solid residuals samples. These data showed (using 100% dry solids) that the WET extraction increased the leached arsenic concentration from 8 to 80 times the leached arsenic concentration in the TCLP method. This increase is expected due to the use of citric acid buffer, a 48-hour extraction period, and anaerobic test conditions for the WET extraction. All these conditions are more aggressive than acetic acid extraction used for TCLP analysis.

The California WET test caused a much higher arsenic release than the TCLP test for the lagoon simulations for almost all of the residuals tested. TCLP arsenic levels tracked much more closely than WET concentrations with lagoon supernatant concentrations. The ratios of WET and TCLP arsenic to measured lagoon arsenic are provided in Table 2.7. The WET test arsenic concentration for the 20% solids concentration residuals was as much as 700 times greater than the lagoon arsenic release. The WET test (100% solids concentration) measured arsenic release was 3 to 7 times higher than the lagoon measured arsenic release. Overall, the WET data, measured using 100% solids, demonstrated a much closer relationship to the lagoon release than did the 20% solids California WET results.

All of these factors emphasize the importance of minimizing residuals quantities generated as a byproduct of arsenic removal processes. An effective arsenic removal process may suffer from regulatory and cost impediments, if residuals from the process are not managed or treated efficiently. The most effective formulations for meeting TCLP and WET standards could be superplasticized cements, which can "fix" arsenic and other metals, especially when the metal concentrations are elevated (US Patent 5,551,976). Gallup (2002) has recently fixed and stabilized an oily sludge containing 9% (w/w) arsenic. The sludge exhibited a TCLP of 1,600 mg/L arsenic to a TCLP of 5 mg/L (the regulatory standard) with a 50-g mortar sample made of 50% solid waste, 25% cement, 25% sand, and 2.5 mL of Sikament 86 superplasticizer (water was added as necessary to make a slurry). Other possible fixatives included magnesium-containing Portland Type I or II cement, fly ash, and clays.

Table 2.6
Settled residuals characterization data

Utility name	Sample description	Solids conc. (percent)	pH	Total As*	Total Fe*	Total Al*	TCLP As (mg/L)	California WET As (mg/L)
Louisiana Water Company (New Iberia), La.	Untreated residuals	4.1	11.5	-	-	-	-	-
	Liquid portion	-	11.3	BDL	460	610	-	-
	Solid portion	100.0	-	110	2,320	11,800	0.012	0.096
City of Great Falls, Mont.	Untreated residuals	3.0	7.5	-	-	-	-	-
	Liquid portion	-	7.3	280	290	430	-	-
	Solid portion	100.0	-	490	14,660	109,160	0.158	4.830
City of Helena, Mont.	Untreated residuals	3.6	6.7	-	-	-	-	-
	Liquid portion	-	7.7	31	3,720	210	-	-
	Solid portion	100.0	-	420	43,730	41,000	0.125	1.880
Dept. of Public Utilities (Billings), Mont.	Untreated residuals†	4.1	6.9	-	-	-	-	-
	Liquid portion	-	7.6	2	1,570	440	-	-
	Solid portion	100.0	-	70	74,670	25,520	0.013	0.863
Lockwood Water Users Association (Lockwood), Mont.	Untreated residuals	4.7	7.2	-	-	-	-	-
	Liquid portion	-	7.4	BDL	26,660	400	-	-
	Solid portion	100.0	-	820	327,700	11,800	0.011	0.224

(continued)

Table 2.6 (Continued)

Utility name	Sample description	Solids conc. (percent)	pH	Total As*	Total Fe*	Total Al*	TCLP As (mg/L)	California WET As (mg/L)
Los Angeles Dept. of Water & Power (LADWP), Calif.	Untreated residuals	14.0	7.2	-	-	-	-	-
	Liquid portion	-	6.8	358	22,010	310	-	-
	Solid portion	100.0	-	730	139,390	16,990	0.162	4.931
Indiana-American Water Company, Ind.	Untreated residuals	1.9	7.2	-	-	-	-	-
	Liquid portion	-	7.7	14	500	210	-	-
	Solid portion	100.0	-	15,730	493,200	1,720	0.031	2.471
Heath, Ohio	Untreated residuals	26	10	-	-	-	-	-
	Liquid portion	-	9.4	BDL	BDL	2.0	-	-
	Solid portion	100.0	-	19	3,300	9,107	BDL	0.697

Data from MacPhee et al. 2003
BDL = Below Detection Level (<2 μg/L for arsenic)
*Units for mixed and liquid samples are μg/L; units for solid samples and mg/kg
†Completely mixed residuals before separation

Table 2.7
Ratio of California WET and TCLP to arsenic lagoon release at two months

	California WET: Lagoon		TCLP: Lagoon	
Utility	20*	100*	20*	100*
New Iberia, La.	63	3.7	5.00	15.8
Great Falls, Mont.	20	3.3	2.30	0.05
Helena, Mont.	8	3.1	0.20	0.14
Billings, Mont.	156	6.6	1.90	0.06
Lockwood, Mont.	362	4.6	1.80	0.35
Los Angeles, Calif.	8	6.9	0.04	0.01
Indiana-American Water Company, Ind.	694	19.0	0.30	0.04

Data from MacPhee et al. 2003
*Refers to percent solids concentration of the residuals used in the test

ARSENIC BRINE SOLIDIFICATION

The handling and disposal of arsenic-laden residuals pose a serious problem because these residuals may release arsenic back to the environment under some disposal conditions. Since the sequestration of arsenic in solids is sensitive to environmental conditions (such as the redox potential, pH, and chemical composition of fluids in contact with the solid phase), changes in environmental conditions, particularly the development of reducing conditions, may promote the release of arsenic from residuals. Changes in environmental conditions may be associated with residual handling and disposal processes, including dewatering, storage, lagooning, landfilling, land application, or discharge to sewer systems. Although there are numerous publications dealing with the treatment of drinking water to remove arsenic, there are relatively few published works that specifically address the characteristics of residuals containing arsenic or the treatment of those residual streams for removal of arsenic (Cornwell, 2001).

Many of the physico-chemical treatment processes used for arsenic removal generate significant quantities of sludge or other solid wastes. In some cases, these solids must be disposed of as hazardous waste, which is subject to strict governmental regulations and poses issues of future liability (Harper and Kingham 1992, Crusberg et al. 1991). The concentration of arsenic in solid residuals is expected to increase as the arsenic concentration in treated potable water is decreased to meet new regulatory standards.

Some arsenic treatment processes, such as ion exchange and RO, generate arsenic-laden liquid brine and concentrate. These brines must be treated to remove arsenic before disposal (i.e., to evaporation ponds or to coastal waters through a brine disposal line). One potential option for removing arsenic from brines is the use of environmentally-benign biopolymeric sorbents. Biosorbents, which can be synthesized from inexpensive natural materials, have been used primarily for removal of heavy metal cations from solutions. Removal of anionic contaminants such as arsenate requires either chemical treatment of the aqueous solution to convert the ions to a cationic species, chemical optimization of the solution to enhance anionic binding to the biosorbent, or modification of the biosorbent to bind anions specifically (Brierley 1990). Compared with coagulant sludge, biosorbents offer increased ease of residual handling and decreased cost of residual disposal.

The use of the biopolymer alginate for removal of cationic metals from solutions has been extensively investigated (Chen et al. 1993). Alginates are available commercially as water-soluble sodium alginates and have been used for more than 65 years in the food and pharmaceutical industries as thickening, emulsifying, film forming, and gelling agents. Alginic acid is a polyuronic acid extracted from various species of brown seaweeds by a multi-step process. A typical extraction process involves first washing and treating seaweed with diluted sulfuric acid (quantitatively transforming calcium alginate into alginic acid), extraction by dissolution with alkali, separation of insolubles, then precipitation as the calcium salt, followed by acid washing to recover alginic acid. The salts of alginic acid are obtained by neutralization. Alginate production is situated in California, the British Isles, Norway, France, Chile, Japan, and China.

Alginic acid is composed of mannuronic (M) and guluronic (G) acid residues. The M and G residues occur in varying proportions depending on the source (Haug et al. 1974) and are arranged in block patterns comprised of homopolymeric regions (MM blocks and GG blocks) interspersed with alternating regions of heteropolymeric regions (MG blocks) (Draget et al. 1998, Fundueanu et al. 1999).

Typically, the gelation of alginate is obtained through the binding of divalent cations, which occurs spontaneously when a sodium alginate solution is added to a solution containing calcium. In this process, interfacial polymerization is instantaneous with precipitation of calcium alginate followed by a more gradual gelation of the interior as calcium ions permeate through the polymer matrix. The resultant gel is biochemically inert and mechanically stable (Velings and Mestdagh 1995). The binding of divalent cations and subsequent gel formation is dependent on the composition and arrangement of the blocks of the residues (Yamagiva et al. 1995). In particular, gel strength is related to G content, which varies from 20 to 70% depending on the seaweed source. The ribbon-like morphology of M blocks is less favorable for cation binding than the coil-like morphology of the G blocks, which provide more favorable cavities for cations (Jang et al. 1990). The carboxyl groups are responsible for the association of metal ions with alginate gels (Schweiger 1962). Divalent cations form junction zones between alginate molecules in what has been termed the "egg box" model (Grant et al. 1973). In this model, the divalent cations are the "eggs" that sit in interstices between parallel alginate chains and interact with carboxyl anion and hydroxyl group ligands on the chains (Figure 2.1). Such structures form only between G segments.

The application of alginate as a sorbent for the removal of arsenic from solution has been investigated by Min and Hering (1998a). They found that the treatment of calcium alginate gel beads with Fe (III) optimizes both sorptive capacity and the physical properties of gel beads. Although Fe (III) alginate beads are easier to produce, their physical properties are less desirable than those of modified calcium-iron alginate (Ca-Fe) beads. Sorption of As (V) by Fe (III) doped calcium alginate gel beads was found to be effective in removing As (V) over a wide concentration range (Min and Hering 1998b). The removal of arsenic from contaminated water by alginate beads loaded with iron oxide has also been investigated (Zouboulis and Katsoyuannis 2002).

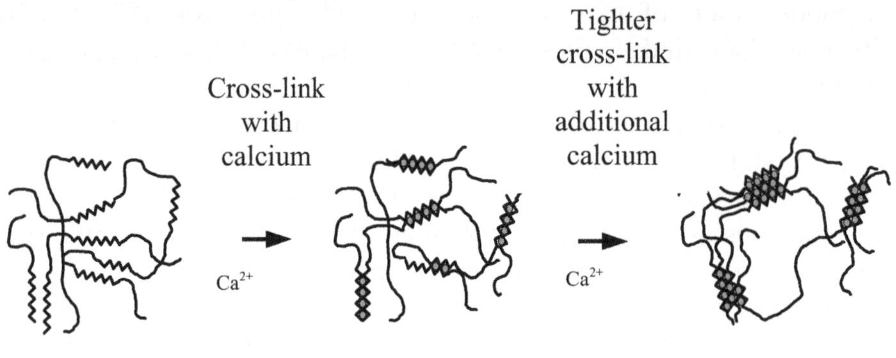

Figure 2.1 Schematic diagram of "egg box" model (Adapted from Grant et al. 1973)

PERCHLORATE AND NITRATE BIOLOGICAL BRINE TREATMENT

Recently there has been much interest in the biodegradation of perchlorate in spent salt brines from groundwater and other treatment process wastewaters. Most of the research has focused on the identification of salt-tolerant bacteria that can reduce either perchlorate or nitrate and on the effects of environmental conditions on the performance of these bacteria.

Woolard and Irvine (1995) evaluated phenol biodegradation in hypersaline wastewater (>3.5% salt) and provided a good summary of factors that must be considered for microbial growth in salt brines. Four basic problems were identified when treating wastes containing 0.1 to 5% salt. First, conventional cultures are sensitive to changes in ionic strength. Shifts in salinity as little as 0.5% can disrupt system performance. Rapid shifts in salinity cause more problems than gradual shifts; however, even systems that have been acclimated to a certain salt content will lose performance if salt concentration is rapidly reduced. Second, increase in salinity generally tends to disrupt normal metabolic function and reduce degradation kinetics. Third, treating saline wastewaters increased the concentration of effluent suspended solids. Forth, the extent of salt acclimation of mixed cultures from typical wastewater treatment systems is limited. A general consensus in literature is that conventional wastewater cultures cannot be used to treat wastes greater than 3 to 5% salt.

To compensate for desiccating osmotic forces present in saline environments, bacterium accumulates compatible solutes to equalize the ionic strength of the cytoplasm with the external environment. Moderate halophiles accumulate a mixture of inorganic cations (Na^+, Ca^+) and organic compounds (amino acids, betaine, glycerol) for osmotic regulation. Many halophiles may require growth factors such as amino acids and vitamins. In this study, it was observed that the addition of iron as an inorganic salt or yeast extract was necessary to prevent nutrient limited growth. It was noted that, in general, nutrient requirements become more complex as salt concentration of the medium is increased.

Okeke, Giblin, and Frankenberger (2002) evaluated salt-tolerant bacteria that could reduce nitrate and perchlorate in salt brine. Microbial cultures investigated were isolated and enriched by multiple transfers. Of the isolates tested in serum bottle tests, the best performer with respect to perchlorate reduction was identified as *Citrobacter sp.* Several environmental parameters were evaluated for their effect on *Citrobacter sp.* performance. Of the nine carbon sources evaluated, yeast extract and sodium acetate gave the best bacterial growth performance. A highly refined, food-grade molasses was tested but did not perform well under the conditions used. Tests were conducted in the 20°C to 40°C range, and 30°C yielded the best performance. Tests were also conducted in a pH range from 6.0 to 10.0, and pH 7.5 demonstrated the best performance. Both denitrifying bacteria and *Citrobacter sp.* growth and perchlorate reduction performance were affected by NaCl concentration of 2.5%. However, *Citrobacter sp.* could grow and reduce perchlorate at 5% NaCl and up to about 7.5% total solids in high nitrate and sulfate brine. In tests conducted with optimized conditions using a co-culture of *Citrobacter sp.* and perchlorate, approximately 50% of the perchlorate was removed in 7 days.

Glass and Silverstein (1999) evaluated denitrification of high nitrate, high salinity, simulated wastewater in a sequencing batch reactor (SBR). Denitrification in salt brines is important because it has been shown that nitrate must be significantly reduced before perchlorate reduction can be effectively completed in a wastewater containing both salts. Results showed that denitrification rate decreased with increasing total dissolved solids (TDS). High concentrations of nitrate (up to 36,000 mg/L) were reduced in a wastewater with TDS up to 18% when the batch reactor system was controlled at pH 9.0. This was achieved by gradually acclimating the activated sludge used in this study, by incrementally increasing TDS and nitrate concentrations from 2,700 mg/L NO_3-N and 4.8% TDS up to 8,200 mg/L NO_3-N and 18% TDS. At a pH of 7.5, denitrification of 5,400 mg/L NO_3-N was completely inhibited.

Logan, Wu, and Unz (2001) collected and enriched six different high-salinity surface water and sediment cultures for the purpose of demonstrating perchlorate reduction in high salinity brine from ion exchange processes. Only two samples exhibited perchlorate reduction at a salinity of 3% or higher. Growth rates of the most promising culture for the Great Salt Lake peaked at 0.06 d^{-1} (a doubling time of 11.2 days) at 5% salinity and showed no growth at 13% salinity. Limited perchlorate reduction results were obtained, but at 3% salinity, perchlorate was reduced from 592 to 45 mg/L after 8 weeks.

Gingras and Batista (2002) isolated and enriched a biological culture from a wastewater treatment plant in the Clark County Sanitation District, near Henderson, Nevada. Lactate was used as a nutrient at mass ratio to perchlorate of 3:1. Increasing the addition of salt from 0 to 1.5%, reduced perchlorate reduction rate over 90% from >11 mg/day to <1 mg/day. The presence of 0.5% salt reduced the growth coefficient by 32%. The conclusion of that day was that this particular mixed culture was not able to biodegrade perchlorate in anticipated ion exchange wastes at acceptable rates.

A review of the Ground-Water Remediation Technologies Analysis Center (GWRTAC) Technology Status Report for Perchlorate Treatment Technologies (Roote 2001) identified few other projects focused on salt brine. The project summaries discussed bench-scale testing and full-scale design work done at Henderson, Nevada (12,000 mg/L TDS), and prototype demonstration at Thiokol (>20,000 mg/L TDS) performed by ARA. Since this GWRTAC report was published, ARA has modified the prototype at Thiokol to increase capacity by over 300% and optimize the process to increase the TDS limit to 30,000 mg/L (Coppola, Rine, and Startzell 2003). The increase in capacity was mostly based on physical plant modifications. The primary modifications were an increase in reactor capacity by 1,800 gallons (or about 75%), and changes in pumps, valves, and flow meters that enable higher flow rates and reduced the system hydraulic residence time from ~28 hours to 10 to 12 hours.

The consortium of microbes used in ARA's perchlorate reduction processes is derived from the consortium containing HAP-1 organism (Attaway and Smith 1993). The HAP-1 organism was isolated and identified as *Wolinella succinogenes*. (Wallace et al. 1996). The microbial consortium containing *Wolinella succinogenes* is known to be salt-tolerant. Sodium and potassium-based brine wastewater, containing perchlorate and nitrate, has been effectively treated by ARA's process at ATK-Thiokol since December 1997 (Coppola, Rine, and Startzell 1998). The salt content of the wastewater at Thiokol has been typically limited to approximately 2%. The ARA patent includes a claim that the ARA process that will biodegrade nitrate and perchlorate in brines containing over 3% NaCl and over 3.4% TDS. (Coppola and McDonald 2000) Recent work by ARA has also shown that this same consortium will biodegrade perchlorate in media up to 7% TDS in batch systems, and up to 4% TDS in flow reactor systems at hydraulic residence times feasible for a commercial process to be economical and practical.

It is interesting to note that *Wolinella succinogenes* was originally called *Vibrio succinogenes* (Bonder et al. 1982). Identification of this organism as being in the genus Vibrio may have been due to the fact that sodium stimulates the growth of most Vibrio species and they are found in aquatic habitats with a wide range of salinities (Holt et al. 1994).

More recently, Gillogly et al. (2002), conducted limited sequencing batch reactor studies with the goal of reducing perchlorate concentration in ion exchange brine (3% NaCl) to less than 200 μg/L. Perchlorate reduction performance was inhibited by dissolved oxygen using acetate as the elector donor. A gradual improvement in performance was observed with time.

Roberts et al. (2002) were not successful at acclimating activated sludge to reduce perchlorate in 3% brine. However, they did have limited success enriching cultures from marine sediments. Batch tests were conducted to evaluate the effect of different nutrients or amendments on the reduction of 100 mg/L of perchlorate. The addition of trace metals and sodium sulfide (Na_2S) resulted in 60% greater perchlorate reduction than did addition of trace metals or sodium sulfite alone.

PERCHLORATE AND NITRATE THERMAL BRINE TREATMENT

The only published data on thermal treatment process for perchlorate destruction in salt brines is the ARA study sponsored by USEPA (Li and Coppola, 2000). The focus of this work was on sodium chloride brines and is based on the principle that perchlorate can be readily decomposed thermally or hydrothermally. Thermal decomposition tests showed that the presence of ferric salts or ferrous salts reduced the decomposition temperature of perchlorate to 170°C. Without the addition of promoting agents, complete thermal decomposition of perchlorate in NaCl brine was obtained at 600°C. Due to the large volume of sodium chloride brine effluent generated by conventional ion exchange processes, use of non-catalyzed, thermal treatment to decompose perchlorate is very energy intensive, which limits this technology's commercial potential. However, regeneration of perchlorate-selective resins (Gu et al. 2000) using the $FeCl_3$/HCl regeneration technique (Gu et al. 2001) has the potential to reduce the volume of brine effluent to 0.2% of the NaCl effluent generated by treating that same quantity of groundwater. Thermal decomposition of ferric chloride brines is much more economical process than thermal treatment of NaCl brines because of the significantly reduced quantity of brine and the catalytic effect of the ferric chloride.

PERCHLORATE AND NITRATE TREATMENT AT A WWTP

Biological processes used to treat high-strength perchlorate wastewater have been utilized since the 1970s. Yakovlev et al. (1973) patented an unaerated batch process to reduce 120 mg/L perchlorate to below 3 mg/L. They used activated sludge from a wastewater treatment plant for the microbial seed and relied on sludge organic matter to provide the necessary substrate. When Korenkov et al. (1976) and Romanenko et al. (1976) seeded their anaerobic bioreactors with Vibrio dechloraticans, they achieved a perchlorate removal rate of 70 mg perchlorate per hour per gram biomass. Ethanol and acetate were added as substrate. Attaway and Smith (1994) patented an anaerobic bioreactor that demonstrated a perchlorate removal rate of approximately 1,500 mg perchlorate per hour per gram biomass. Aged brewer's yeast and a mixed-culture containing *Wolinella succinogenes* served as the substrate and the microbial seed, respectively.

While a number of researchers have used isolated culture from a wastewater treatment plant sludge to develop biological technologies for treating high strength perchlorate-laden wastewater (Hurley et al. 1996, Wallace et al. 1998), the literature review indicates that the treatment of perchlorate at a municipal wastewater treatment plant has not been demonstrated. Biological treatment processes designed to remove nitrate are a common component of municipal wastewater plants, and these processes include suspended growth systems, fixed-bed systems, rotating biological contactors, and fluidized-bed systems. Naturally occurring bacteria provide nitrate-reducing biomass, and substrate requirements are met with organic matter present in the wastewater or by adding an exogenous carbon source such as methanol (Tchobanoglous and Burton 1991). Given that many denitrifying bacteria also have perchlorate-reducing capabilities (Coates et al. 1999, Rikken et al. 1996), biological wastewater systems designed for nitrate removal would also likely be effective for perchlorate removal.

the microbial seed and relied on sludge organic matter to provide the necessary substrate. When Korenkov, et al. (1976) and Romanenko et al. (1976) seeded their anaerobic bioreactors with Vibrio dechloraticans, they achieved a perchlorate removal rate of 70 mg perchlorate per hour per gram biomass. Ethanol and acetate were used as substrate. [illegible] (1994) patented an anaerobic bioreactor [illegible]

[illegible]

CHAPTER 3
ARSENIC BACKWASH WATER MINIMIZATION AND RECYCLE

INTRODUCTION AND OBJECTIVES

Pre-filtering the raw groundwater source is often used to decrease backwash frequency for various treatment systems that use adsorption media, which are used for arsenic treatment in drinking water. The typical arsenic concentration in the backwash water from these processes is not well characterized, and thus, discharge and recycling the backwash warrants evaluation for full scale implementation of such systems. Some existing plants using media filters do incorporate a prefiltration step (e.g., some AA plants in New Hampshire), but recycling the backwash of such processes, especially using replaceable filters, has not yet been investigated in detail, and would thus be an innovative approach. Such alternatives will further decrease the production of on-site arsenic waste discharge. The main objective of this task is thus to minimize backwash water to reduce total amount of waste generated during adsorber operation. In the bench-scale plan, source water with arsenic concentration greater than 20 μg/L were filtered using cartridge filter, media, or bag filter to achieve this objective.

The raw groundwater was collected from the participating utilities in six 5-gallon (gal) jugs for prefilter evaluation. Testing parameters included headloss accumulation (if any observable in the experimental timeframe), total filterable residue, turbidity, arsenic (total and soluble), iron (total and soluble), manganese (total and soluble), and particle counts for both raw and pre-filtered water.

Another possible method for minimizing the volume of residuals is recycling treated backwash water to the head of the plant as mentioned. The objective of this task is to reduce waste generated during operation by treating backwash water and recycling the backwash water stream to develop a "near-zero liquid discharge" option. Backwash water treatment methods considered consisted of bag filtration and coagulation/sedimentation (bench-scale) and cartridge filtration (pilot-scale).

BENCH-SCALE TESTS MATERIALS AND METHODS

Total filterable residue (also known as total suspended solids (TSS)) was analyzed using Standard Method 2540 D, with Whatman glass filter GF/A, with nominal pore size of 1.6 μm. Minimum detectable dried residue yield should be 2.5 mg per sample volume. Particle counting was performed using Standard Method 2560 D (Light scattering) with minimum particle size of 2 μm and minimum concentration of 10 counts/mL (or 1,000 counts/100 mL). The range of particle sizes measured was: 2, 3, 5, 7, 10, 15, and >15 μm. In order to suspend settled particles, the sample was gently swirled with a glass rod for 10 seconds. Turbidity measurements were done by Standard Method 2130 B (nephelometric) with Hach turbidimeter (detection limit of 0.01 NTU). For arsenic, iron, and manganese, inductively coupled plasma (ICP, EPA Method 200.7, Standard Method 3120B) with detection limits of 2, 10, and 5 μg/L, respectively, was used. The difference between total and soluble was determined using a 0.45 μm filter. EPA 200.7 was used for raw, backwash, filtered, as well as extracted arsenic from California WET, which is described in Chapter 4. Although EPA Method 200.8 (ICP-MS) allows for lower detection limits (DL), and therefore has become the method to use under the arsenic rule, the DL of EPA Method 200.7 was deemed sufficient for the sample analysis of the bench-scale and stabilization tests.

The bench-scale experimental setup for the cartridge filter included a peristaltic pump and a volumetric flow rate indicator. The kit included all parts necessary to perform constant flow rate testing. For each nominal pore size of the filter products (5 to 10 μm), 1-inch long cartridge test segments were used (MicroKlean III, Cuno, Meriden, Conn.). The flow rate of the peristaltic pump was set at 416 mL/min to provide about 2 gpm/ft^2 or 1.1 gpm/EQSL (equivalent single length is the number of 10-inch long cartridges to be used). This flow, although below that recommended by the manufacturer (3 to 5 gpm/EQSL), was selected because it corresponded to a reasonable filtration rate for sand and thus could form a basis for performance comparison.

A peristaltic pump was adjusted to load at about 2 gpm/ft^2 a 0.5-inch diameter PVC filter column containing 12-inches of sand with an effective size (d_{10}) of approximately 0.5 mm and with a uniformity coefficient of 1.3 (d_{60}/d_{10}). After ten bed volumes were filtered, a filtered water sample was collected and analyzed. The filter was backwashed after each experiment.

Unlike cartridge filter manufacturers, bag filter manufacturers do not generally have specific bench-scale protocols that generate useful information with a minimal quantity of water. Therefore, standard felt polyester bag filters with nominal pore sizes of 5, 10, and 50 μm (PE-5-P-12, PE-10-P-12, and PE-50-P-12, respectively, Rosedale Products, Ann Arbor, Mich.) were tested by pumping raw groundwater with a peristaltic pump at a flowrate of 140 mL/min through a 47-mm diameter filter holder that contained a disc section cut out from the various bag filters. This flowrate corresponds to the same filtration rate of 2 gpm/ft^2 as for the other prefiltration devices, but is at the low end of the normal filtration range for bag filters.

BENCH-SCALE TESTS RESULTS AND DISCUSSION

The groundwater samples that underwent bench-scale prefiltration experiments (for backwash minimization) came from four different utilities: Stockton, California; Many Farms, Arizona; Mesa, Arizona; and Lakewood, California The latter city also served as the site for the pilot plant. Groundwater characteristics are illustrated in Table 3.1. The results from the bench-scale study on arsenic, turbidity, and particles for backwash minimization techniques are presented in Figures 3.1 to 3.3. In all cases, soluble arsenic concentration was only slightly below that of total arsenic. The difference between total arsenic and soluble arsenic after prefiltration may be owed to small arsenic particulates passing through the filter. A difference in soluble arsenic levels is due to arsenic retained by the prefilter, which would make its landfill

Table 3.1
Groundwater characteristics used for arsenic filter tests

Parameter	Stockton	Lakewood	Mesa	Many farms
Turbidity (NTU)	1.29	0.45	0.42	0.50
TSS (mg/L)	4.5	0.6	0.4	3.6
Particles (#/mL)	3,200	410	910	660
pH	7.9	7.9	7.6	7.1
As (μg/L) – total (soluble)	44 (28)	34 (29)	32 (28)	51 (51)
Fe (mg/L) - total	0.62	0.36	BDL	0.27
Fe to As ratio (total weight)	14:1	11:1	NA	5:1

Values in parentheses denote soluble As concentration. Soluble Fe was below detection limit (BDL).

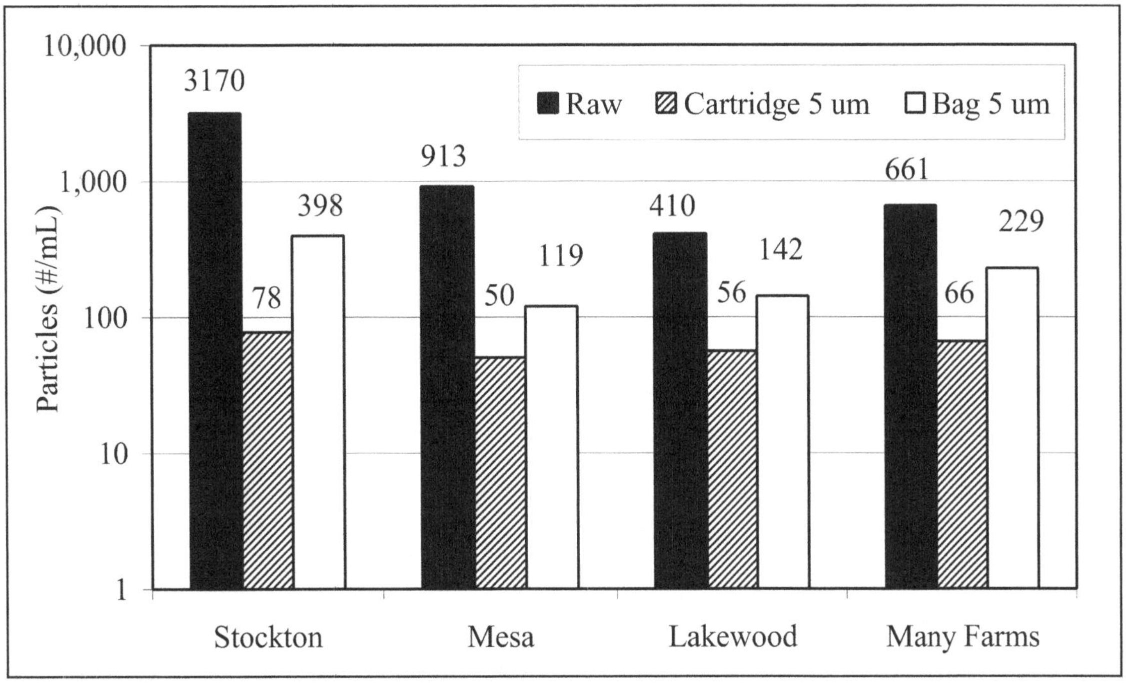

Figure 3.1 Comparison of particle removal levels achieved by cartridge and bag filters at same nominal pore size (5 µm) for groundwater samples from various sites

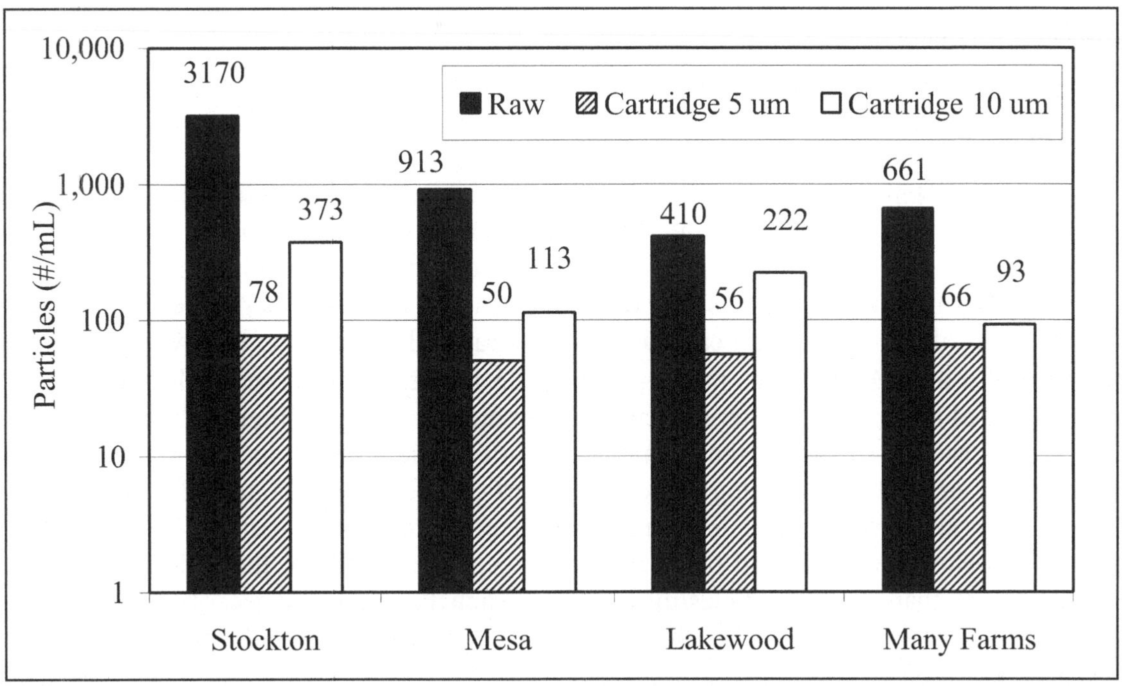

Figure 3.2 Comparison of particle removal levels achieved by cartridge filters with increasing nominal pore size (5 to 10 µm) for groundwater samples from various sites.

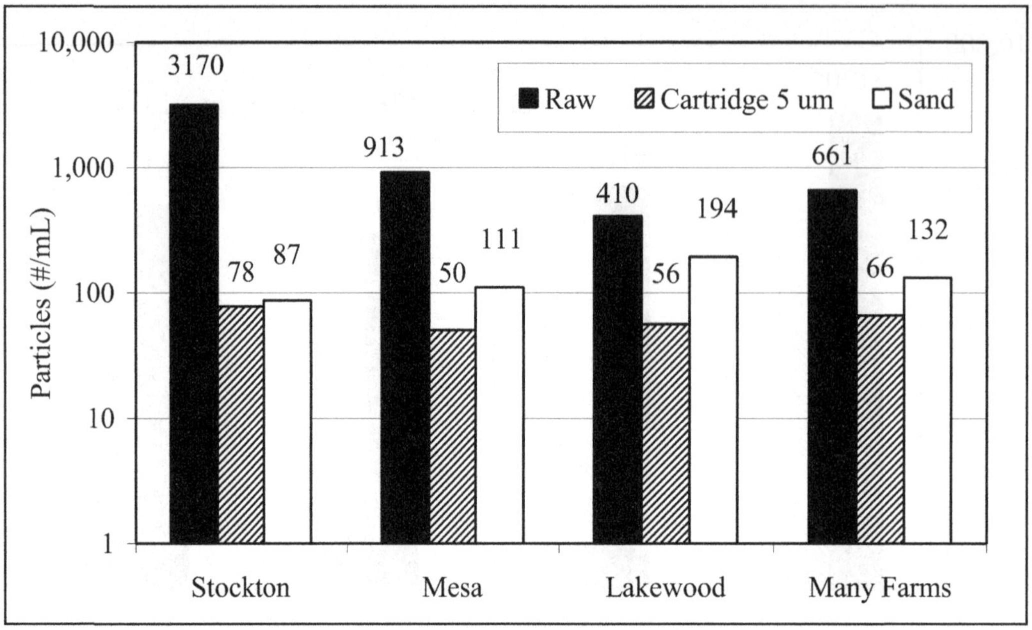

Figure 3.3 Comparison of particle removal levels achieved by sand and 5-μm cartridge filter for groundwater samples from various sites

disposal classification problematic. Since all samples were from groundwater, the water was very clear (mostly under 1 NTU) and analyzing the prefiltered water for TSS would have required most of the water sample. Thus TSS analysis was not performed. Also not presented are the results for dissolved iron for all groundwaters because the concentrations were either below detection limit (BDL) or very low (in general < 0.3 mg/L for total iron). In addition, sample collection made it difficult to maintain conditions similar to that of time sampling and resulted in dissolved iron to be oxidized.

Soluble arsenic concentration remained constant after all prefiltration devices at about 30 μg/L for Stockton, Lakewood and Mesa and approximately 51 μg/L for Many Farms, confirming that arsenic did not adsorb onto the various media (cartridge, bag, or sand). The particles results in Figure 3.1 shows that for equal nominal pore size, cartridge filters performed better than bag filters. The 10-μm cartridge demonstrates close to 1-log removal in particles for Stockton, Mesa and Many Farms (>1-, ~1- and 0.85-log removal, respectively) suggesting that most particles of these groundwater sources were greater than 10 μm (Figure 3.2).

In the case of Lakewood, the 5-μm bag, 10-μm cartridge filter, and the sand all performed similarly with only about 0.35-log removal of particles (Figures 3.1 to 3.3). Only the 5-μm cartridge filter achieved about 1-log removal indicating that most particles were in the 5 to 10 μm range. However, it should be noted that except for Stockton, the groundwater samples were clear (less than 1,000 particles/mL), and as such the performance of prefilters remains relative to the particle levels in the raw water (i.e., log removals would be higher with more turbid water).

Sand performance was consistently between that of the 5- and 10-μm cartridge filters. In terms of performance of the backwash minimization techniques for the various groundwater sources, the 5-μm cartridge was consistent in providing an absolute value for the prefiltered water of less than 100 particles/mL. The 10-μm cartridge filter and 5-μm bag filter performed about the same regardless of groundwater source.

Arsenic Backwash Recycle

Participating utilities collected about 40 gals in eight 5-gal jugs of spent backwash water from GFH or E-33 adsorbers. Testing parameters included headloss accumulation on the bag filter, TSS, turbidity, arsenic (total and soluble), particle counts, iron (total and soluble), and manganese (total and soluble) for both untreated and treated spent backwash water using the method described previously.

The bag filtration test consisted of pumping the spent backwash water with a peristaltic pump through a 47-mm diameter filter holder that contained a disc cut out from various bag filters (Rosedale, Ann Arbor, Mich.). A range of filtration rates for each bag filter type was used at each nominal pore size (from 90 mL/min corresponding to about 5 gpm/ft^2, to 350 mL/min corresponding to 20 gpm/ft^2) to cover the typical backwash rate experienced at the plants (generally 10 gpm/ft^2). Rosedale Products list loading rates for bag filters up to 45 gpm/ft^2. Each test lasted about an hour. For the pilot plant, an equalization basin was placed prior to filtration to uncouple the bag filtration rate from the backwash rate, which allows control of the recycle flow (i.e., the recycle flow could theoretically be as low as it takes to empty the equalization basin before it gets refilled with the next backwash).

The coagulation and sedimentation tests were conducted using a jar test apparatus consisting of up to six 2-L jars and Phipps & Bird variable speed mixers. Best results were duplicated for reproducibility. The ferric chloride dose range was based on earlier work (MacPhee et al. 2003) by the project team and optimized during this project. During ferric chloride experiments, the coagulation time was 20 min and the settling time was assessed by drawing water 4 in. (10 cm) below the 2-L water mark at 2.5, 3.5, 5, 10, and 25 min, representing 1.0, 0.75, 0.5, 0.25, and 0.1 gpm/ft^2, respectively. Sedimentation overflow rates covered an order of magnitude in a range comparable to full-scale clarification facilities. Results for coagulation/sedimentation are shown for an equivalent sedimentation rate of 0.25 gpm/ft^2. The $FeCl_3$ dose shown in the results corresponds to the dosage for which optimal turbidity removal was achieved for the specific spent backwash water sample.

The backwash samples came from three sites, two of which had also submitted groundwater samples. Mesa, Arizona was operating two adsorption systems: GFH from US Filter and E-33, an iron-based proprietary sorbent (Bayoxide) from Severn Trent. Both sorbents have three times as much absorption capacity in terms of bed volumes as activated alumina (Driehaus et al. 1998). Therefore, these promising techniques will appear attractive to utilities, and it is important to consider the implications of backwashing these processes. Whereas headloss buildup from accumulated particles may have corresponded to media exhaustion for activated alumina, the increased adsorption capacity of GFH and E-33 suggest that backwashing for headloss mitigation will take place prior to media exhaustion. An approach to treat this spent backwash to standards acceptable for recycling would provide much-needed water conservation in arid areas where arsenic is prevalent. Such an approach would also minimize discharging issues associated with arsenic residual.

Parameters to assess the performance of the various techniques for treating backwash intended for recycle included arsenic, iron, particle count, turbidity, and headloss buildup. Headloss was measured from the pressure gauge at regular time intervals. It generally increased linearly until a plateau was reached, or when no flow passed through the filter. For the pilot plant tests, unacceptable headloss occurred when the differential pressure was greater than specified by the filter manufacturer. It is important to return a treated backwash stream with an arsenic concentration less than or equal to that of the raw water to prevent accumulation of this inorganic contaminant. Monitoring for iron and particle levels also served to assess removal performance with respect to other water quality parameters. Most iron detected in the untreated spent backwash is the result of GFH or E-33 sorbent particles being carried over during the backwash sequence. Therefore, it is expected that any arsenic detected in the backwash would be predominantly adsorbed to such particles, and as such a decrease in arsenic when using a filtration technique (i.e., bag filters) would remove these iron particles. Table 3.2 presents the untreated backwash characteristics for Bakersfield, California (GFH); Mesa, Arizona (GFH and E-33); and Many Farms, Arizona (GFH). The pH of the backwash samples was about neutral to slightly alkaline. Based on Table 3.2, the mass ratio of Fe:As was 76 for Mesa GFH, about 78 for Mesa E-33, and approximately 85 for Many Farms. The backwash sample from Bakersfield had a mass ratio of Fe:As of 400.

Figure 3.4 shows that bag filtration with nominal pore size of 10 μm or less removed total arsenic by at least one order of magnitude. However, this removal performance was not repeated with the other spent backwash samples, in particular for Many Farms. This is because soluble arsenic comprised a greater portion of the total arsenic for Many Farms (24 μg/L out of 33 μg/L) and Mesa E-33 (38 μg/L out of 97 μg/L) than for Bakersfield (3 μg/L out of 86 μg/L). Therefore, the total arsenic detected in the filtered backwash corresponds approximately to the soluble arsenic level in the untreated spent backwash. This also means that soluble arsenic did not adsorb onto the bag filter.

A significant decrease in total iron, with a level almost matching that of the dissolved iron level in the untreated backwash is shown in Figure 3.5. These results confirm the hypothesis that particulate (i.e., non-soluble) arsenic detected in the backwash and removed by bag filtration was adsorbed to sorbent particles carried over from the backwash. Once adsorbed to GFH, arsenic does not desorb during a physical backwash even when some of the sorbent is carried over. This is an important factor to consider for residuals management aspect. Since raw water is used as backwash source, the maximum level of soluble arsenic in spent backwash treated by bag filtration will be the level of soluble arsenic in the raw water. Thus, soluble arsenic concentration in the treated backwash may be less than the maximum level (i.e., soluble arsenic concentration in the raw water) based on the degree of GFH exhaustion and backwash rate.

Table 3.2
Backwash characteristics

Parameter	Stockton	Lakewood	Mesa	Many farms
Turbidity (NTU)	10.2	4.2	6.8	2.1
TSS (mg/L)	59.5	-	14.4	4.6
Particles (# /mL)	11,300	6,760	10,000	11,360
As (μg/L) – Total (Soluble)	86 (3)	30 (28)	97 (38)	33 (24)
Fe (mg/L) – Total (Soluble)	34.4 (0.25)	2.28 (BDL)	7.51 (0.03)	2.79 (0.17)

Values in parentheses denote soluble As concentration.
BDL = below detection limit.

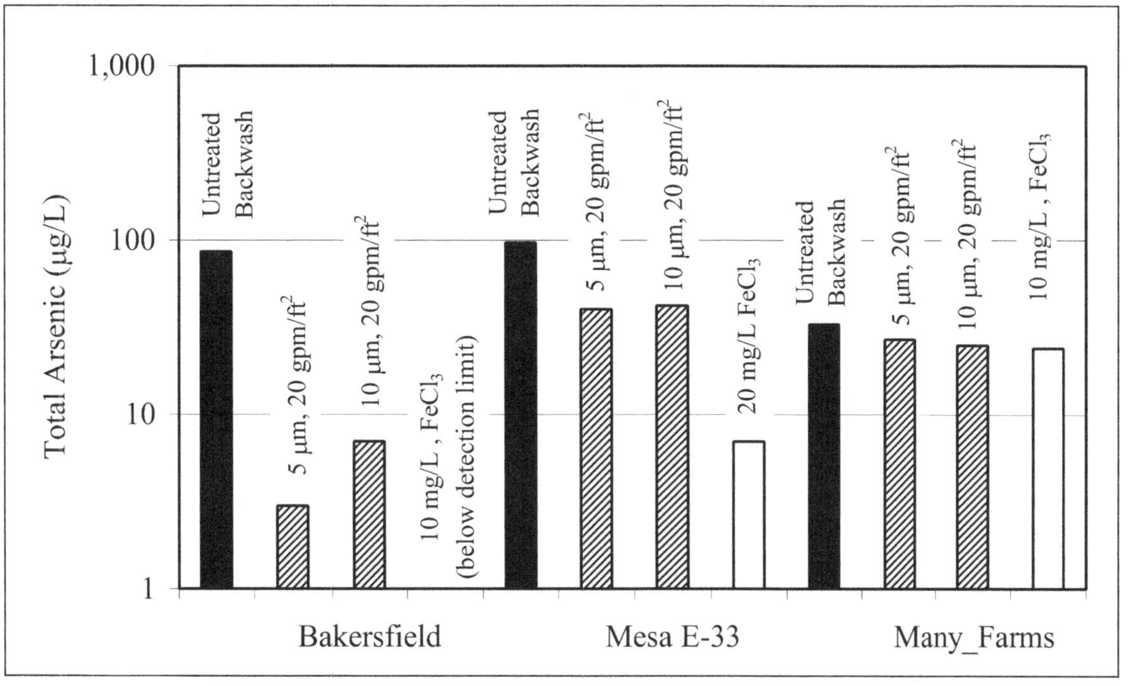

Figure 3.4 Arsenic removal in spent backwash water by bag filtration and coagulation-sedimentation for Bakersfield, Mesa E-33, and Many Farms

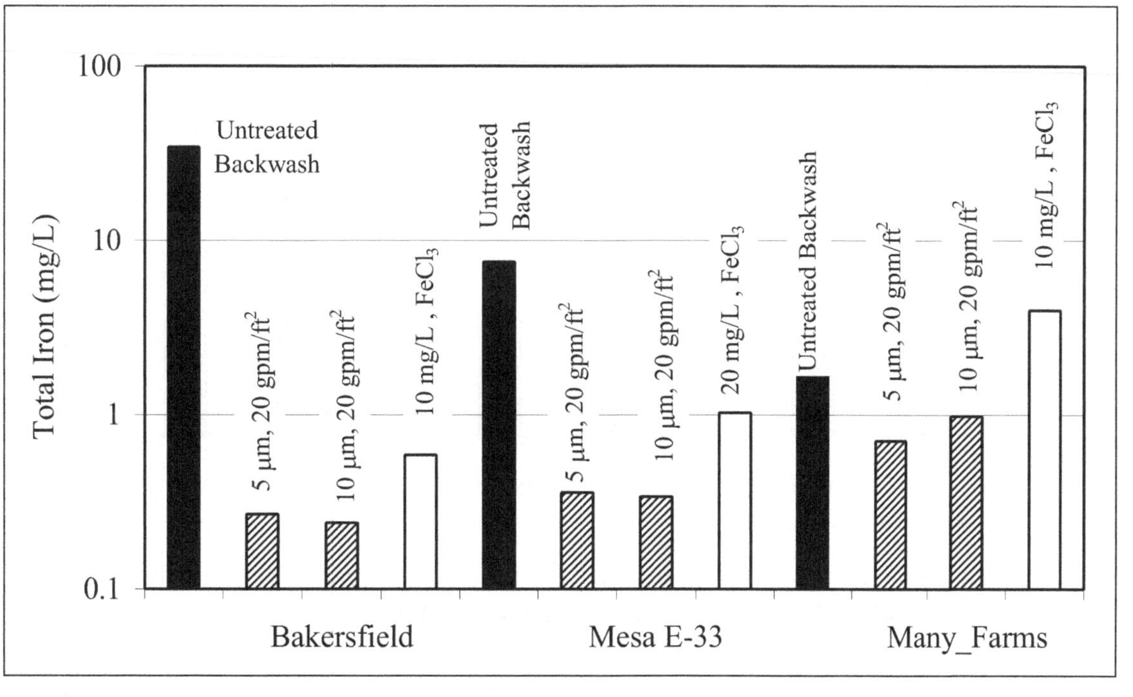

Figure 3.5 Iron removal in spent backwash water by bag filtration and coagulation-sedimentation for Bakersfield, Mesa E-33, and Many Farms

Coagulation sedimentation consistently outperformed bag filtration for the removal of turbidity (Figure 3.6), particles (Figure 3.7), and arsenic. The addition of sufficient amount of ferric chloride coagulant resulted in the adsorption of particulate as well as dissolved arsenic. As a result, the total arsenic in the treated backwash was lowered to below that of the soluble arsenic level of the untreated spent backwash. Although great results were achieved in using ferric chloride, the total iron level in the treated backwash was higher compared to that of the bag filter (Figure 3.5) due to unprecipitated ferric chloride from potential overdosing.

Since it was shown that arsenic and iron removal by bag filtration depends on the particulate form of these constituents, the effect of combining different pore sizes and filtration rates was focused on particle removal. Figure 3.8 shows that for Mesa's GFH spent backwash, a lower filtration rate through the bag filter appears preferable in terms of particle removals, with the 10-μm bag filter at 5 gpm/ft^2 outperforming the 5-μm bag filter at 10 gpm/ft^2. The reason for filtration rate having a greater impact than nominal pore size may be due to particle being large enough to offer no distinction between a 5-μm and a 10-μm bag filter. To prevent accumulation of these particles in the feedwater when recycling the treated backwash, it is thus important to consider the nominal pore size of the bag filter and the filtration rate. It should be noted, however, that the coagulation and sedimentation method performed better than bag filters.

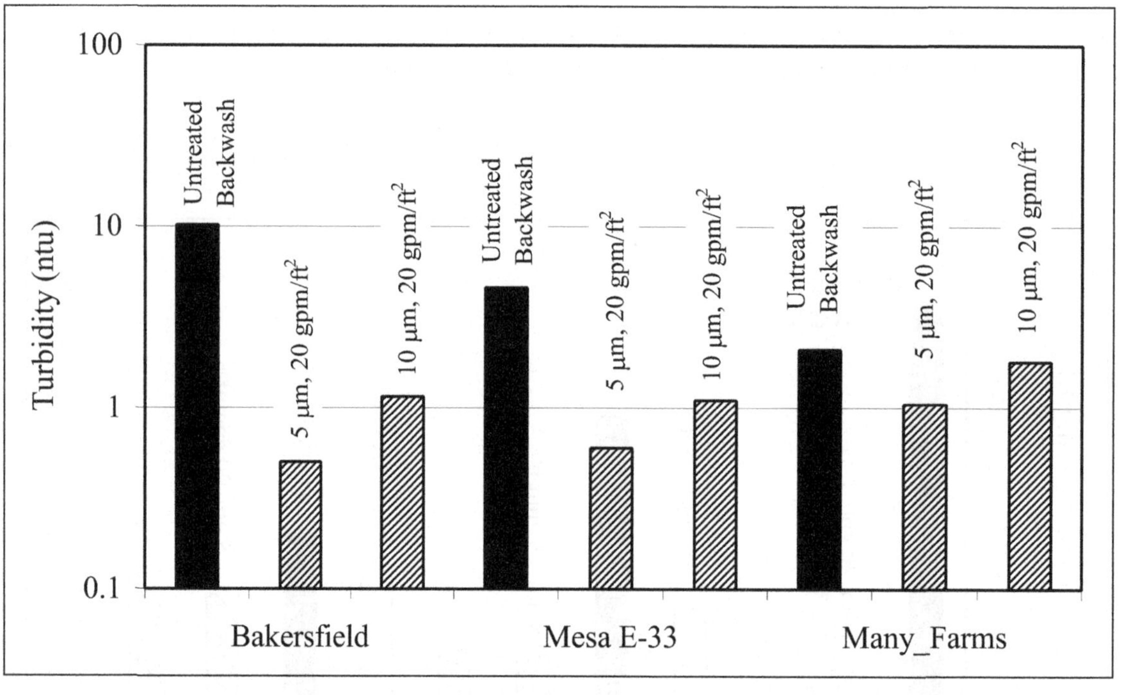

Figure 3.6 Turbidity removal in spent backwash water by bag filtration and coagulation-sedimentation for Bakersfield, Mesa E-33, and Many Farms

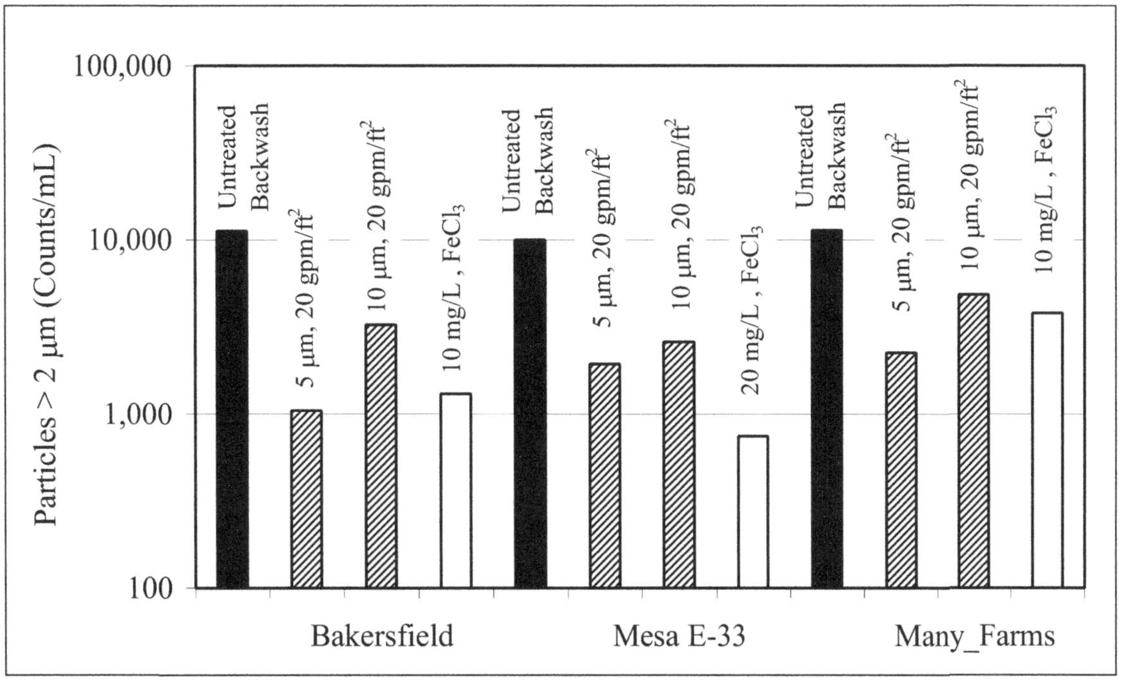

Figure 3.7 Particle removal in spent backwash water by bag filtration and coagulation-sedimentation for Bakersfield, Mesa E-33, and Many Farms

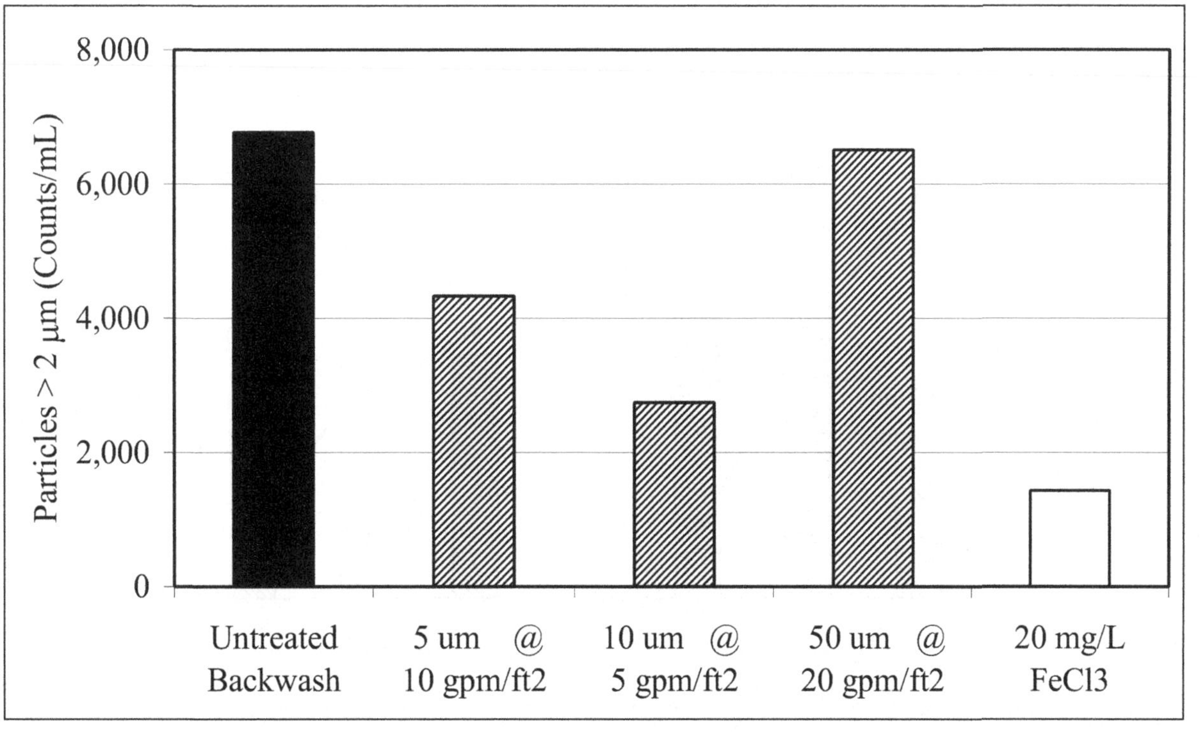

Figure 3.8 Particle removal in the GFH spent backwash sample of Mesa, Ariz. by coagulation-sedimentation and by bag filtration of various pore sizes and filtration rates

Filtration rate was found to be a factor in particle removal for spent backwash samples from Bakersfield, but the impact was minimal for Mesa E-33, as shown in Figure 3.9. For Bakersfield's spent backwash (GFH), the lower filtration rate (5 gpm/ft^2) and smaller nominal pore size resulted in optimal operating conditions since acceptable particle removal was achieved for minimal headloss development. For Mesa's E-33 spent backwash, varying the filtration rates did not have any significant impact on water quality for the 5 µm bag filters. At the high filtration rate of 20 gpm/ft^2 the 5-µm bag filter performed better than the 10-µm bag filters, which indicate that nominal pore size is a greater factor than filtration rate, contrary to Bakersfield's GFH spent backwash. Based on Figures 3.8 and 3.9, filtration rate and nominal pore size have a direct impact on headloss buildup and it generally appears that operating at the lowest filtration rate and smallest nominal pore size maximized particle removal while significantly minimizing headloss buildup.

PILOT-SCALE FIELD DEMONSTRATION MATERIALS AND METHODS

The pilot system was set up to assess the backwash minimization and the backwash recycle techniques concurrently. Figure 3.10 is a process diagram of the pilot plant, and Figures 3.11 through 3.12 are pictures of the pilot plant installed at Well No. 6 in Bloomfield Park, Lakewood, California. The pilot plant consisted of two treatment trains, each with its own GFH vessel. One train was dedicated to backwash minimization and as such included (ahead of the GFH vessel) a 10-µm resin bonded cartridge filter (MicroKlean III from Cuno, Meriden, Conn.) of the same type as used in the bench-scale tests.

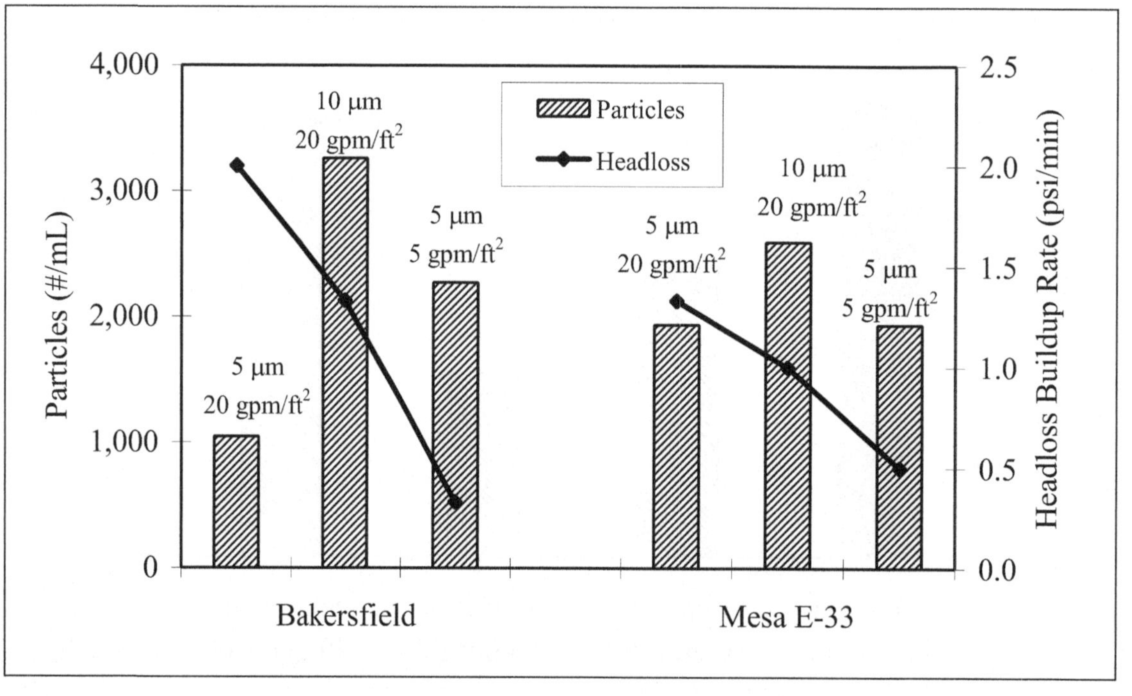

Figure 3.9 Particle count and headloss buildup rate for treatment of backwash samples from Bakersfield and Mesa E-33 by bag filtration

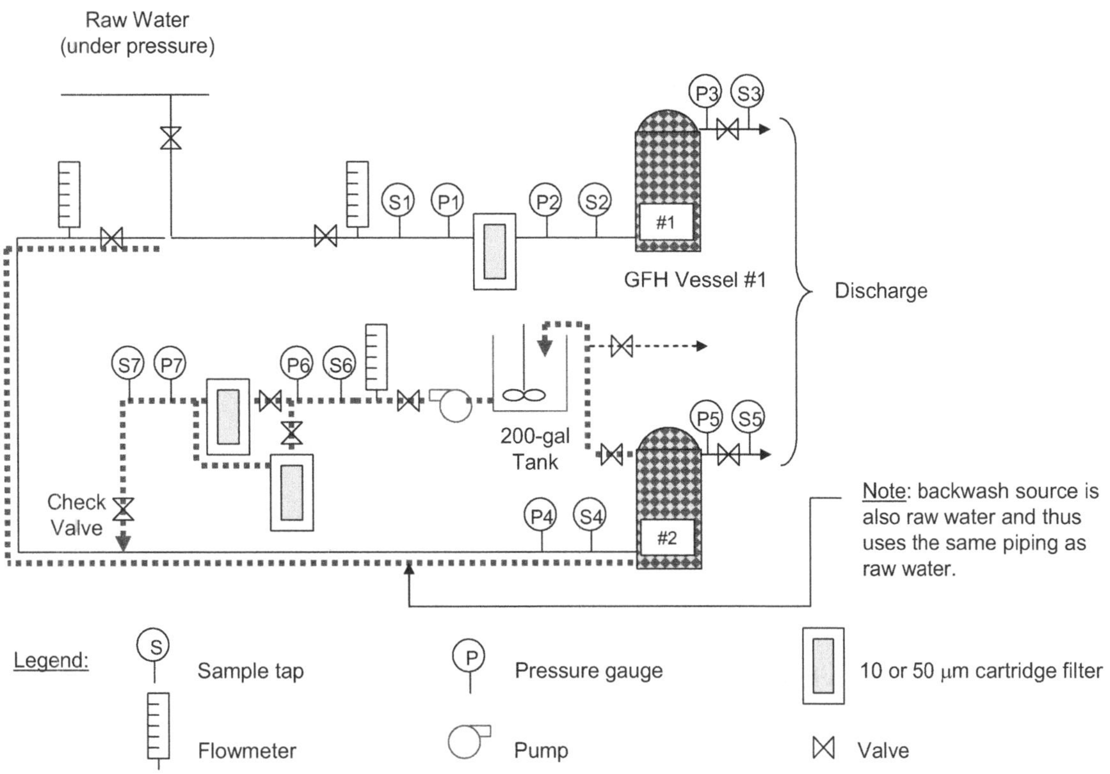

Figure 3.10 Process schematic of pilot plant for backwash minimization and backwash recycle in Lakewood, California

Figure 3.11 General view of the GFH pilot plant with backwash holding tank in Lakewood, California

Figure 3.12 Close up view of the backwash minimization train (right side) and the backwash recycle train (left side)

The other train was dedicated to the backwash recycle and included two resin-bonded cartridge filters (MicroKlean III from Cuno, Meriden, Connecticut) in the backwash discharge line of the GFH vessel, a 200-gal barrel to receive the treated backwash, and a small positive displacement pump to inject the treated backwash back into the feed line. With two filters different filtration rates were tested, as both could be used concurrently. Cartridge filters were used in place of bag filters for treating the spent backwash for the following reasons:

- Lakewood groundwater was found to be the least turbid among all other groundwaters, and headloss from accumulated particles would develop more slowly across the GFH. Hence, any backwash initiated within the normal frequency will be less turbid than the backwash tested in the bench-scale experiments (probably less than 10 NTU). Under such conditions, the necessity of bag filters for handling highly turbid water was not as important, since resin bonded cartridge filters can operate well with turbidities even around 10 NTU.
- 5 μm was found to be the optimal nominal pore size for the bag filters in the backwash recycle bench-scale tests, and the bench scale minimization tests showed that particles were removed to about the same extent for the 5-μm bag filters and 10-μm cartridge filters.
- Having the same kind of filters with the same nominal pore size for backwash minimization and backwash recycle can allow for a straight comparison of the effectiveness of either method in particle removal.
- Cartridge filters are more versatile than bag filters and are more readily available than bag filters for different flows (e.g., cartridge holders, etc.).

Since the treatment train dedicated to backwash does not have a prefilter prior to the GFH vessel, it can serve as control in comparison to the backwash minimization train when no backwash is initiated or being recycled. Daily pressure readings (P1 through P5) and turbidity sampling (S1 through S5) were carried out by the utility operators. The loading rate for each GFH vessel was about 5 gpm/ft^2 based on manufacturer recommendation. For each sample tap, daily samples taken during normal train operation (i.e., no backwash event taking place) were combined into a composite sample for arsenic and iron analysis.

For each backwash event, initiation, treatment, and recycle of the backwash took place within a day. The backwash recycle flow was set to empty the equalization tank within 8 hours resulting in a recycle flow of about 8% of the feed flow (this recycle flow could be further minimized if backwash recycle was to take place over 24 hours or longer). Four backwash events, each at the manufacturer-recommended backwash rate of 10 gpm/ft^2, were conducted to test the effect of filtration rate and nominal pore size. The four pilot testing conditions for backwash treatment and recycle consisted of two 10-μm cartridge filters in parallel, two 50-μm cartridge filters in parallel, a single 10-μm cartridge filter, and a single 50-μm cartridge filter. For each of these backwash events, pressure readings (P3 through P7) and turbidity samples (S4 through S7) were collected. One sample for total and dissolved arsenic and iron was collected at sample tap S6, a composite sample was used for determining total arsenic and iron at S7, and one sample was used to measure total arsenic at S4 as shown in Figure 3.10.

PILOT-SCALE TESTS RESULTS AND DISCUSSION

Total arsenic in the raw water was measured at 30 μg/L, with 27 μg/L in the dissolved form, which is consistent with the sample that was used in the bench-scale experiments. As in the bench-scale backwash minimization tests, arsenic essentially passed through the cartridge filter, indicated by a total arsenic concentration of 28 μg/L at S2 (cartridge filter effluent). Since most of the arsenic (about 90%) was in the dissolved form, the cartridge filter material does not statistically (i.e., allowing for analytical variations) sorb arsenic, and therefore the cartridge is not subjected to hazardous waste classification.

Pressure and turbidity were measured before and after the GFH vessels and cartridge filters to assess whether prefiltration is significantly beneficial to GFH operation (i.e., if particles retained by the cartridge filter result in significantly delaying the onset of backwash). Figure 3.13 plots the differential pressure progression (i.e., the differential pressure rate of increase each day – psi/day) across the cartridge filter, across GFH No. 1, and across GFH No. 2 for three sequential experimental runs lasting 5, 6, and 4 days respectively, for the normal cycle (i.e., just running the groundwater through the cartridge filter and GFH adsorbers). The raw source water remained essentially the same throughout the runs in terms of total arsenic, iron, and turbidity levels. The headloss buildup across the cartridge filter was two to four times that of the headloss buildup across GFH No.1. The headloss progression across GFH No. 2 was consistently less than across GFH No. 1. The latter result may be attributed to uneven initial conditions in both adsorbers in terms of particle loading, although both were backwashed prior to the tests. These results also showed that the cartridge filter did not decrease the headloss buildup across the GFH vessel. Although the cartridge filter retained particles greater than 10 μm as indicated by the turbidity removal in Figure 3.14 and by the rapid headloss buildup, headloss minimization on GFH No. 1, if any, was not significant.

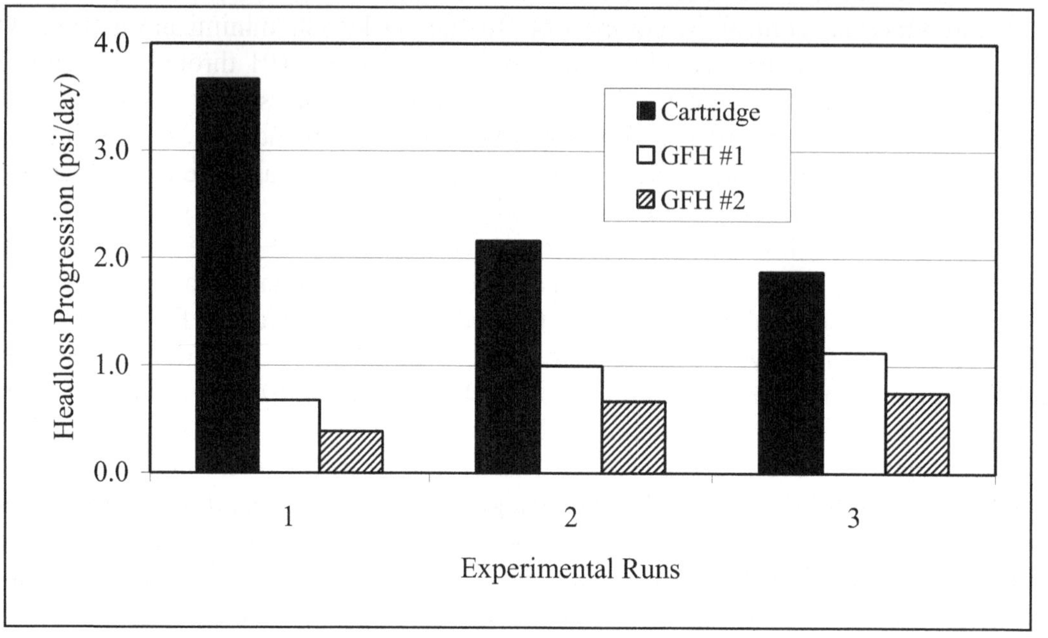

Figure 3.13. Headloss buildup rate (psi/day) across the 10-µm cartridge filter and GFH vessels (normal cycle)

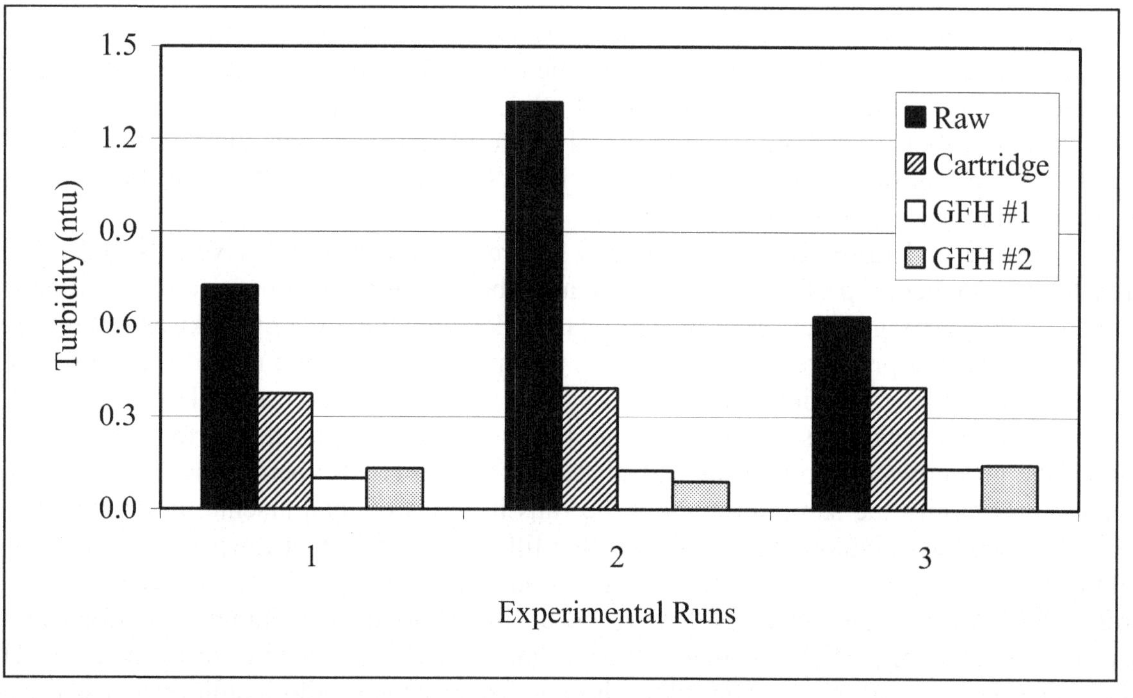

Figure 3.14. Turbidity levels in the raw water and in the effluent of the-10-µm cartridge filter and GFH vessels (normal cycle)

It is possible that particles greater than 10 μm do not comprise the bulk of the groundwater particles and are not significantly contributing to any potential headloss buildup on the GFH. Moreover, during each test the cartridge filter color turned from white to dark orange. Since raw water was supplied from a tap typically unused in the pressure tank, the flow could have contained unsettled sediments that were subsequently trapped by the cartridge filter. The fact that headloss progression across the cartridge filter for each successive experimental run was lower than for the previous one also suggests such contribution by the aged pressure tank.

Additionally, since the filtration area of the cartridge filter is much smaller than that of the adsorber, the effect of particle loading significantly reduced run time to less than a week. The differential pressure across the cartridge filter reached 10 to 15 psi in a few days. Consequently, the duration of the tests may have been too short to observe any significant trend for headloss development across the GFH adsorbers. However, the flow through the single cartridge filter (about 5 gpm) was consistent with the manufacturer-recommended filtration rate of 3 to 5 gpm/EQSL. It is probable that additional cartridge filters would attenuate the headloss buildup across the filters to a point where any potential benefit on the GFH vessel could be justified (e.g., a month instead of a week).

For all backwash events, and throughout the recycle duration, the differential pressure across GFH vessel No. 2 remained at about 3 psi (i.e., there was no increase in differential pressure throughout the day), which shows that recycling the filtered backwash did not result in accumulated headloss for the adsorber. This could be due to the low flow of the recycle stream (8% of the influent flow), the quality of the returned stream, or a combination of both.

The untreated backwash turbidity ranged from about 200 to 400 NTU, which is about 10 to 20 times the anticipated turbidity of the spent backwash. Figure 3.15 compares the effect of nominal pore size (10 μm and 50 μm) on turbidity removal in the beginning and towards the end of the backwash recycle test. For both nominal pore sizes the decline in turbidity removal is severe. The reduction was from 90% to 60% for the 10-μm cartridge filter, and from 60% to 10% for the 50-μm cartridge filter. The cartridge filters appeared to have rapidly saturated during the backwash treatment. Having two cartridge filters in parallel compared to a single one did not alleviate the decline in turbidity removal.

Figure 3.16 shows the turbidity resulting from combining the treated backwash stream with the influent and recycle flow and the impact on the GFH effluent turbidity towards the end of the recycle test. Since the raw water turbidity over the course of the entire pilot testing was approximately 1 NTU, Figure 3.16 demonstrates that recycling the treated backwash increased the turbidity in the influent to the adsorber (almost an order of magnitude in the case of the 50-μm cartridge filter), yet the impact on GFH effluent turbidity remained minimal since it only increased 0.1 to 0.2 NTU from the level encountered during normal operation (0.1 to 0.2 NTU). The GFH vessel had therefore the capacity to assimilate the increased influent turbidity without a significant change in headloss buildup or effluent turbidity.

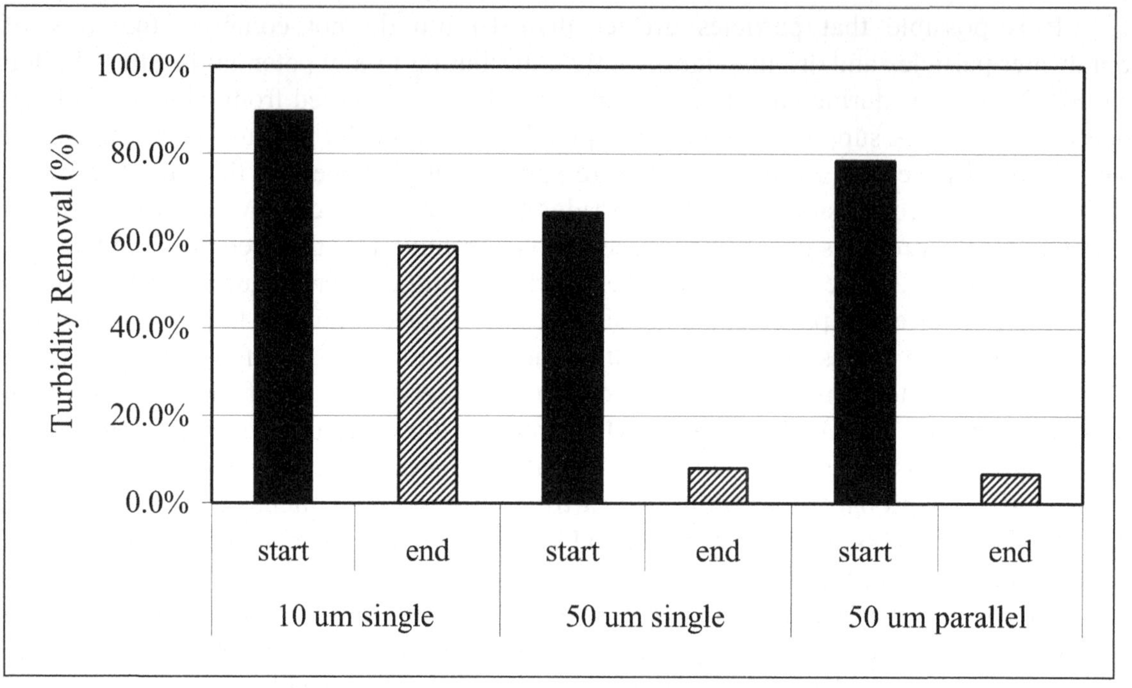

Figure 3.15 Turbidity removal achieved at start and end of a pilot backwash recycle event for single 10-µm and 50-µm cartridge filters, and for two 50-µm cartridge filters in parallel

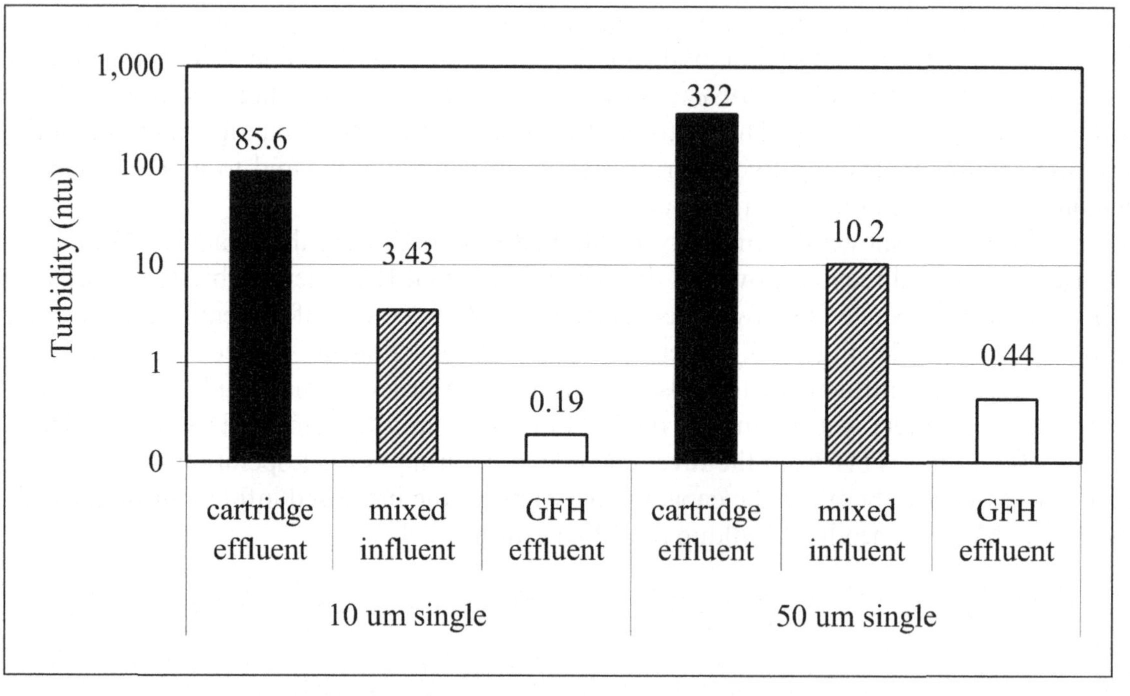

Figure 3.16 Effect of cartridge effluent on turbidity of mixed influent and GFH effluent at end of pilot backwash recycle event

Total arsenic and total iron concentration in the untreated backwash reached as high as 640 µg/L and 67 mg/L, respectively, due to GFH particle carryover. Soluble arsenic concentration in the untreated backwash was on average about 27 µg/L, with 95% of the arsenic is in the particulate form adsorbed to the iron hydroxide particles. The soluble arsenic concentration is also slightly below that of the soluble arsenic concentration in the raw water (about 30 µg/L). A small portion of soluble arsenic in the raw water used for backwash becomes adsorbed to GFH particles during backwash, and that physical backwash does not result in desorption of arsenic previously adsorbed to the ferric hydroxide particles. Figure 3.17 shows the removal of total arsenic and total iron by 10-µm and 50-µm cartridge filters. The 50-µm cartridge filter removed about 30% less arsenic and iron than the 10-µm cartridge filter. For both nominal pore sizes, iron removal was 50% greater than arsenic removal. The apparent discrepancy in the removal may be due to different degrees of arsenic adsorption on the GFH particles, with arsenic preferentially adsorbing on the smaller GFH particles.

Since there is still a significant amount of particulate arsenic passing through the cartridge filter, the mixed influent (i.e., the combination of raw water and recycle streams) is affected. Depending on the nominal pore size of the cartridge, the total arsenic concentration in the mixed influent increased by about 10 to 50%, as shown in Figure 3.18. The total arsenic concentration in the GFH effluent was affected in the same proportion. Although the GFH adsorber attenuated turbidity peaks, it was sensitive to arsenic concentration peaks in the influent. Such result is shown in each backwash recycle event, which caused a significant increase in total arsenic concentration in the GFH effluent as illustrated in Figure 3.19. During the normal cycles, the arsenic concentration in the GFH effluent remained relatively stable.

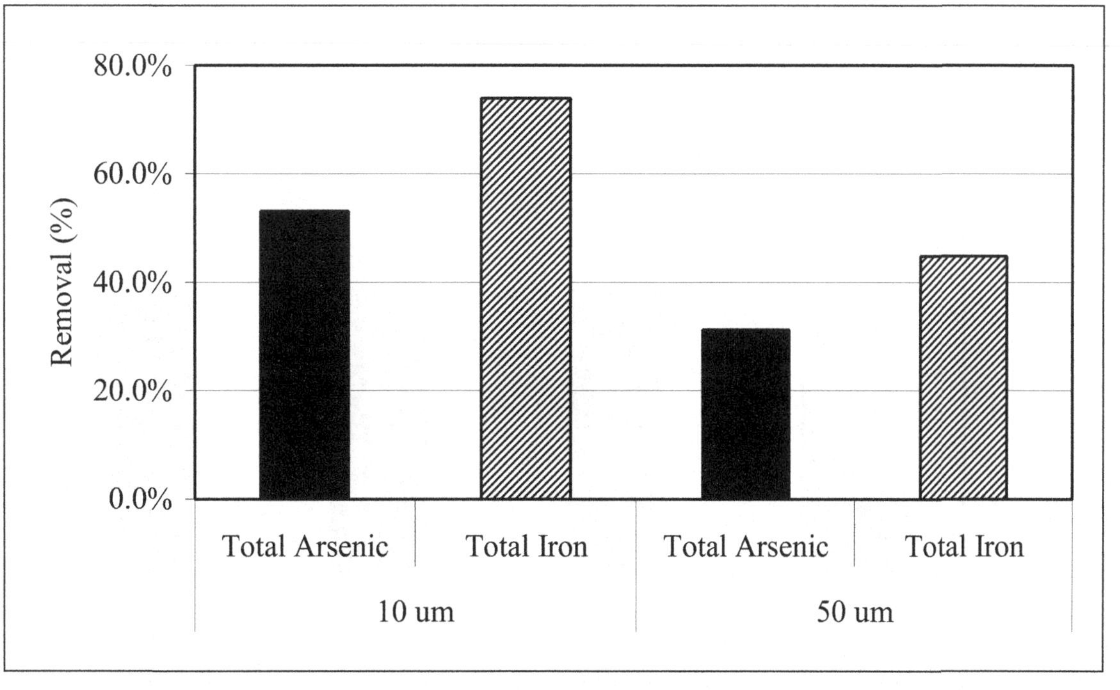

Figure 3.17 Comparison of total arsenic and total iron removal levels achieved by 10-µm cartridge filter and 50-µm cartridge filter during backwash treatment in pilot plant

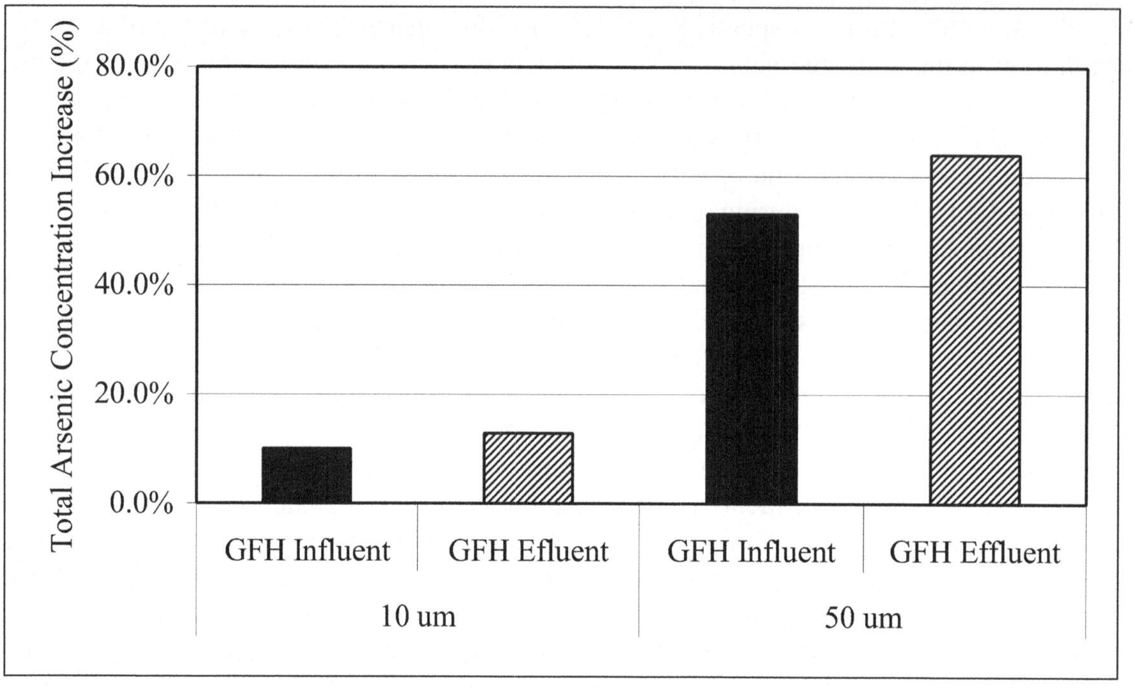

Figure 3.18 Concentration increase in pilot plant GFH influent and effluent due to recycle of backwash filtered by 10-μm cartridge filter and 50 μm cartridge filter

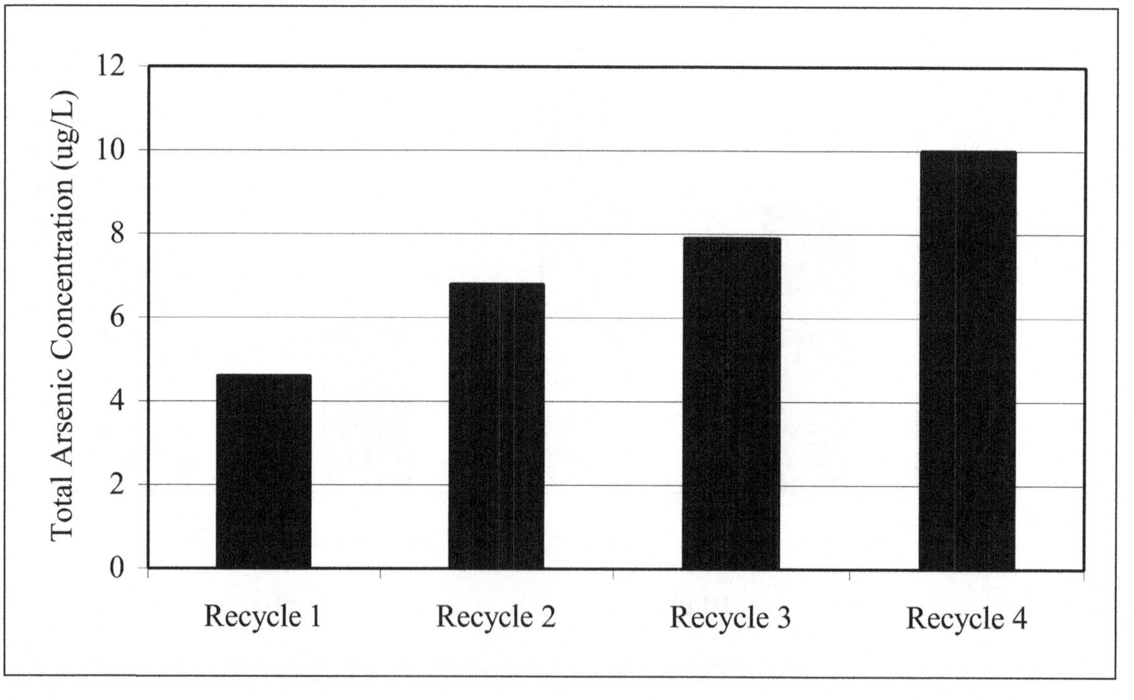

Figure 3.19 Total arsenic concentration in GFH effluent after each backwash recycle event

Based on the pilot test results, prefiltration appears to require a significantly higher number of cartridge filters than was pilot tested (i.e., a significantly smaller filtration rate per cartridge filter than the typical value of 3 to 5 gpm/EQSL) to prevent particle loading in the GFH without causing a rapid headloss increase across the cartridge filters. Backwashing the pilot GFH vessel clearly resulted in significant GFH particle carryover demonstrated by the elevated turbidities in the spent backwash (200 to 400 NTU), although the manufacturer-recommended backwash rate was used (10 gpm/ft^2). Under these conditions, recycling could significantly impact GFH performance, unless the iron hydroxide particles (and adsorbed arsenic) are sufficiently retained during backwash treatment. This in turn would potentially cause a disposal problem for such cartridges, depending on TCLP or California WET results for arsenic adsorbed on GFH. The backwash events showed that cartridge filters were not appropriate in successfully retaining a sufficient quantity of carried over GFH particles. This would mean that unless backwash flow could be controlled as to eliminate particle carryover, an alternative backwash treatment technique, such as coagulation-sedimentation, should be considered.

SUMMARY AND RECOMMENDATIONS

Backwash minimization by introducing a prefiltration step to trap particles that would otherwise contribute to headloss buildup in the adsorber, or treatment and recycle of spent backwash were studied for their ability in reducing the total volume of waste generated during adsorber operation and therefore increasing treatment recovery. The recycle option would also have the benefit of eliminating discharge issues for backwash.

Bench-scale and pilot tests have shown that cartridge filters did not adsorb soluble arsenic, and that particulate arsenic, which was partially retained by cartridge filters, resulted from arsenic adsorption onto iron particles. Thus, disposal of cartridge filters in landfill may not be an issue, depending on TCLP or California WET results for arsenic bound to sorbent particles. Regarding backwash minimization, pilot tests demonstrated that filtration rate through the cartridge filter(s) would need to be lower than manufacturer-recommended (3 to 5 gpm/EQSL), although no headloss buildup had been detected in the bench-scale tests. This result underlines that while bench-scale tests can be an excellent screening tool, they do not replace the value of extended pilot testing.

For backwash treatment and recycle, the turbidity of the backwash samples used in the bench-scale tests was at least an order of magnitude lower than the turbidity of the spent backwash in the pilot tests. Since the backwash samples were composite samples, it is reasonable to assume that most of the turbidity had been captured. Therefore, the high turbidity of the spent backwash could be due to GFH media particles that were carried over, although the manufacturer-recommended backwash rate of 10 gpm/ft^2 was used. The hypothesis of GFH media carryover is substantiated by the elevated total iron concentration (up to 67 mg/L) in the spent backwash. This strong discrepancy in the backwash turbidity between the bench and pilot tests indicates that the cartridge filters did not perform as anticipated. Backwash anomalies may occur in real-world operation, and the pilot tests show that an alternative backwash treatment technique, such as coagulation-sedimentation, would be better suited.

Although untested at the pilot-scale for lack of suitable sites, coagulation-sedimentation performed well in the bench-scale tests by removing more particles than cartridge filters and by also decreasing the total arsenic concentration in the spent backwash to levels lower than that of the groundwater. However, this technique would be more operator-intensive than cartridge or bag filtration, and would imply high capital costs for utilities without such existing facilities (which is often the case when the source water is groundwater). Such approach also requires further treatment of generated residuals, especially if TCLP or other tests indicated arsenic leaching from the settled sludge.

It is recommended that the cost of providing a prefiltration step be compared with the cost of treating the spent backwash based on the results from this study. Further study of the coagulation-sedimentation process for backwash water at a larger scale is also recommended as the most viable option given that support equipment can be provided.

CHAPTER 4
ARSENIC SLUDGE STABILIZATION

INTRODUCTION AND OBJECTIVES

Removal of arsenic by chemical precipitation generates arsenic containing sludge. The objective of this task was to evaluate stabilization methods to minimize arsenic leaching from the sludge. One method involved the addition of lime (MacPhee et al. 2003) for stabilization while an innovative approach was to use a method to include sand, cement, and a superplasticizer to fix the arsenic into a mortar (Gallup 2002). From previous work (MacPhee et al. 2003), when the LADWP ferric chloride sludge was stabilized with lime, only a small amount of arsenic was leached using the TCLP method (USEPA Method 1311). Under the California WET (CA SOP 910), which is more aggressive than TCLP (see Table 4.1), the leached arsenic was above the maximum contaminant level (5.0 mg/L). Therefore, for this research project only the WET test was used to compare both stabilization methods.

Generally, the mixture in Gallup (2002) aimed at a proportion of 50% solid waste, 25% sand and 25% cement to achieve acceptable fixation into a mortar, with such materials as commercially available Portland Type I/II cement and Sikament 86 superplasticizer. Since, the ultimate destination of the stabilized sludge in this project is a landfill (i.e., unlike Gallup's intent, which was to use actual bricks for construction of roads or dikes), the tests with the cement and superplasticizer were modified to allow a direct comparison between lime and cement, and lime and cement with superplasticizer (i.e., quantities of cement and sand were much lower than those required for a mortar).

MATERIALS AND METHODS

Ferric sludge samples from LADWP were used for the bench scale stabilization tests. Samples were collected on June 24, 2003 from backwash pond No. 4 that was out of service and was being prepared to be cleaned by transferring the residuals into piles, so they can be more easily removed by truck. The sample used for the California WET procedures consisted of residual taken from one of the piles. This sample was moist (approximately 45% solids content) and may have its origin near the bottom of the pond. Lime or cement, with or without 5 mL superplasticizer, was added to the residuals in different proportions to form 50-g stabilized

Table 4.1
Key differences in TCLP and California WET extraction procedures

Parameter	TCLP	California WET
Extraction fluid	Glacial acetic acid	Citric acid
Extraction period	18 hours	48 hours
Anaerobic conditions	No	Yes, by purging with N_2 gas prior to agitation
pH	4.9	5.0
Liquid: solid ratio	20:1	10:1

samples containing 10%, 20%, and 50% of fixative by weight (see Table 4.2). Water was added as needed to make a slurry. Samples were left to cure for one week. The untreated sludge was analyzed for total iron, total arsenic, and pH. Additionally, WET procedure was performed on an untreated sludge sample for control.

For this test, 25 grams of each sample was placed in the extraction vessel along with 250 mL of the sodium citrate extraction solution. Nitrogen gas was flushed through air scrubbers placed in the mixture for 15 minutes to remove oxygen from the extraction medium. During the 48-hour extraction period, the sample was vigorously agitated. Extracts were filtered and acidified before analysis.

RESULTS AND DISCUSSION

Portland cement is a complex mix of many compounds affecting the hydration and chemical characteristics of the cement. It is manufactured commercially by heating together a mixture of limestone and clay up to a temperature of 1,300 to 1,500°C. Although 20% to 30% of the mix becomes molten during the process, the majority of the reactions taking place are solid state in nature and therefore tend to be slow. Once cooled, the resulting clinker is ground to a fine powder and a small amount of gypsum (calcium sulfate dihydrate) is added to slow down the rate at which the cement hydrates to a workable level. Most Portland cement clinkers contain principally the following compounds: tricalcium silicate ($3CaO{\cdot}SiO_2$), aluminate ($3CaO{\cdot}Al_2O_3$) and a ferrite phase from the ($2CaO{\cdot}Fe_2O_3$ - $6CaO{\cdot}2Al_2O_3{\cdot}Fe_2O_3$) solid solution series that at one time was considered to have the fixed composition ($4CaO{\cdot}Al_2O_3{\cdot}Fe_2O_3$). Lime is a white crystalline solid with a melting point of 2,572°C. It is manufactured by heating limestone, coral, sea shells, or chalk, which are mainly $CaCO_3$, to drive off carbon dioxide.

Table 4.2
Arsenic release results from California WET test

Sample	Sludge (g)	Lime (g)	Cement (g)	Sikament (mL)	Leached arsenic (mg/L)
1	50				19.6
2	40	10			3.39
3	45	5			5.42
4	25	25			0.18
5	40		10		1.50
6	45		5		5.77
7	25		25		1.31
8	40		10	5	2.27
9	45		5	5	6.46
10	25		25	5	1.20

The raw sludge had a pH of 6.7 and contained 75,000 mg/kg of arsenic and 867,000 mg/kg of iron, resulting in a Fe:As ratio (w/w) of slightly more than 11:1. These concentrations are about five times higher than those analyzed in a previous report (MacPhee et al 2003). Due to the method the ferric chloride residuals are accumulated and concentrated in the piles, the arsenic and iron concentrations are not homogeneous throughout the pile depth. Consequently, residuals sampled at various depths yield various iron and arsenic concentrations. Speculation on the reason for the discrepancy remains, however, outside the scope of this work. Results of arsenic release from different ratios of sludge and additives are presented in Table 4.2 and Figure 4.1.

As the ratio of lime, cement, or cement with superplasticizer to sludge increases, the release of arsenic decreases. For samples with 10% fixative (by weight), arsenic releases were above 5 mg/L, ranging from 5.42 to 6.46 mg/L in the case of lime and cement associated with Sikament, respectively. With 20% fixative (by weight), all arsenic releases were below 5 mg/L and cement was the most effective of the additives in limiting arsenic release (only 1.31 mg/L). For 50% fixative (by weight), lime was the most effective in limiting arsenic release with 0.18 mg/L. Cement with Sikament increased arsenic release by about 0.7 and 0.75 mg/L compared to cement only, when the fixative quantity represented 10% and 20% (by weight) of the stabilized sludge, respectively. When the quantity of fixative increased to 50% (by weight) of the stabilized sludge, the quantity of released arsenic was essentially the same for cement with Sikament and for cement only. Thus, the superplasticizer effect was minimized in the cement/Sikament fixative when the cement quantity was increased to 50% (by weight) of the stabilized sludge (the superplasticizer quantity remained constant at 5 mL as cement amounts were increased, see Table 4.2). Gallup (2000) had obtained a significant reduction in arsenic release from an oily sludge exhibiting a TCLP of about 1,600 mg/L by using a mortar made of 50% residuals, 25% cement, 25% sand, and 2.5 mL Sikament, as discussed in Chapter 2. The difference in the superplasticizer effectiveness with this study may be due to the type of arsenic leachability test employed (i.e., TCLP versus California WET). It is also possible that no significant further reduction in arsenic release was observed since arsenic leaching from the untreated residual is already low.

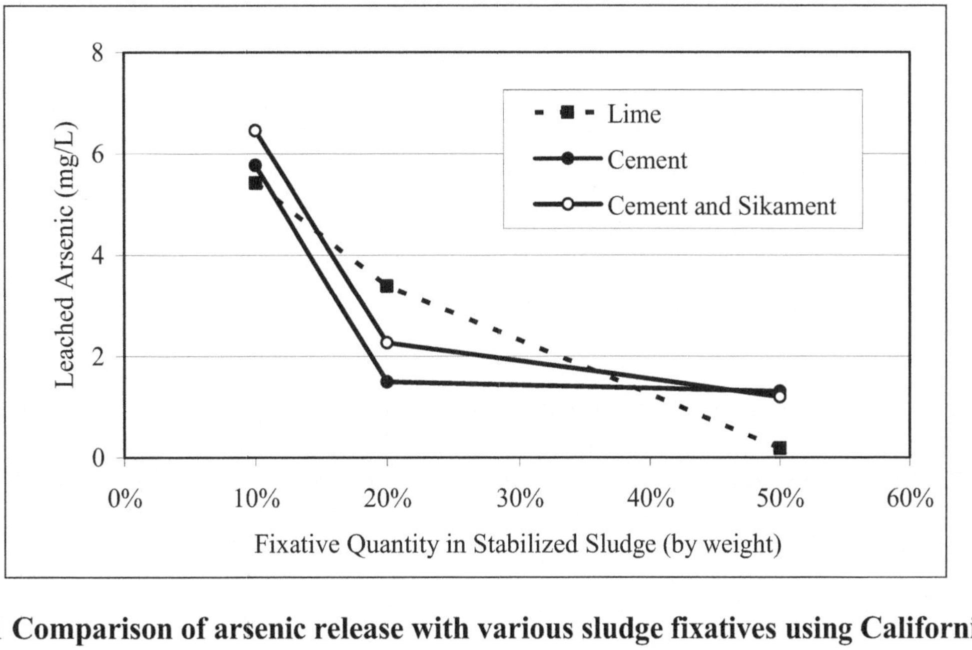

Figure 4.1 Comparison of arsenic release with various sludge fixatives using California WET

The quantity of lime in the conditioned ferric sludge in Parks et al. (2003) ranged from about 1% to 20% (by weight), and they attributed the immobilization of arsenic in the presence of lime at pH<11.5 to the neutralization of the surface charge on the hydroxide solids by soluble calcium rather than to the formation of calcium arsenate solid. However, the amount of lime in the stabilized sludge in our project ranged from 10% to 50%, which means that at pH>12 the probability of calcium arsenate precipitation increases. This hypothesis explains the near linear proportion in arsenic immobilization as the lime concentration in the conditioned sludge increases.

Jing et al. (2003) studied the immobilization mechanisms of arsenate in ferric sludge stabilized with cement. The cement:sludge ratio ranged from 2.5:1 to 20:1 (w/w), indicating that the quantity of cement added was at least two orders of magnitude greater that the range used in this project. The author concluded that in their case, cement treatment significantly reduced arsenic mobility because of the formation of the sparingly soluble calcium arsenate. Indeed, the large cement quantities in that study increased the pH level to about 11.5, whereas the pH in this project was probably slightly alkaline due to the lower quantities of cement added. Since lime has a greater proportion of calcium than cement, it seems likely that for an equivalent ratio on a weight-by-weight basis lime would perform better than cement at a pH where calcium arsenate solid could form. It is possible that the immobilization mechanism be due to surface complexation when limited amount of cement is used. The plateau observed in arsenic release for cement amounts greater than 20% (by weight) could be a transitional phase before the conditions for calcium arsenate precipitation take place.

SUMMARY AND RECOMMENDATIONS

Based on California WET conditions, arsenic leaching from LAWDP's ferric chloride residuals amended with 10% lime or cement (weight basis) decreased by 70% (from 19.6 mg/L to less than 6 mg/L). However, the quantity of fixative need to be increased to approximately 15% to meet the WET regulatory limit of 5 mg/L of leached arsenic. Currently the regulatory limits of TCLP and WET are set at 5 mg/L for arsenic, so the residuals meeting the California WET limit would also comply with the TCLP limit, since test conditions are more stringent for WET. Based on the data from MacPhee (2003), WET results were about two orders of magnitude greater than TCLP results for coagulation residuals treated with lime, and as such a California WET limit of 5 mg/L would still be more stringent than a TCLP limit of 1 mg/L. It would be difficult to extrapolate the actual amount of fixative needed to meet a potential TCLP limit of 1 mg/L from the WET data presented here. The cost of landfilling hazardous materials (including any special hauling requirements) needs to be compared to the cost of adding sufficient lime to meet the regulatory limit for landfilling non-hazardous materials as well as the cost of increased weight of the residuals. In either case, the residuals need to be in a solid form with no free flowing liquids, as defined by the Paint Filter Liquids Test (EPA Method 9095) for compliance with 40 CFR 264.314 and 265.314.

The results from this study and previous research (MacPhee et al. 2001) suggest that lime be used for stabilizing arsenic-containing residuals from coagulation sedimentation processes rather than cement (with or without superplasticizer). Lime is more cost competitive than cement, and the water industry is already familiar with lime additions of up to 20% (w/w) and using filter presses for dewatering water treatment plant residuals. Furthermore, elevated lime concentrations in the stabilized sludge may valuable to the farming community if the stabilized residuals are land applied.

CHAPTER 5
ARSENIC BRINE SOLIDIFICATION

INTRODUCTION AND OBJECTIVES

This project task focused on the use of a biopolymer-based hydrogel sorbent to treat arsenic-containing brines. Previously, chemical precipitation has been typically used to treat such brines. One alternative, as described in Chapter 2, is use of iron-doped, biopolymer-based hydrogel sorbent to selectively remove arsenic in brine. Although the technology has been applied for arsenic, selenium, and chromium(VI) in wastewater, it has not been used to treat concentrate or brine. One of the innovative aspects of this technology is in dehydration and solidification of the spent media. The sorbent exhibits significant volume and weight reduction upon dehydration to minimize residuals produced.

Tests were conducted with As(V) since As(III) is not retained by ion-exchange resins and would therefore not be concentrated in the brines. Alginate beads were formed by cross-linking the polymer in a calcium solution and then doped with iron(III) (Ca-Fe gel) to promote arsenic uptake. It was intended that the spent beads were dehydrated after use as a "single-pass" sorbent for arsenic.

The specific objectives of the work were to: 1) determine the physico-chemical stability of the Ca-Fe doped alginate gel beads in brine solutions; 2) determine the effects of high ionic strength and specific competing ions (i.e., phosphate and sulfate) on As(V) uptake by the Ca-Fe gel beads in batch tests; 3) assess the stability of dehydrated spent beads in leaching tests following TCLP and California WET protocols; and 4) assess the performance of the Ca-Fe gel beads in a column as compared to the batch systems. In addition, some preliminary work was conducted with a modified gel bead in which hydrous ferric oxide (HFO) was incorporated as the sorbent phase within the gel matrix.

MATERIALS AND METHODS

Biopolymer Handling and Storage

Sodium alginates with different gulurionic (G) and mannuronic (M) content were procured from various sources (Sigma Co., FMC BioPolymer, ISP Alginates and Wako Chemicals). An appropriate amount (from 2 to 4 g) of sodium alginate powders were transferred to a 100 mL volumetric flask containing Milli-Q water. A long-stem polypropylene funnel was used to avoid contact of powder with the glass since it did not come off easily from the glass when the water was added. The solution was left on a shaker overnight or until no gel mass was observed. The shaker was used to dissolve the polymer powder for all concentration ranges and viscosity grade powders. Sonication was not used for the dissolution step, because it has been reported (Martinsen et al. 1989) that this process can break the chain, thus changing the characteristics of the polymer. Depending on the concentration and viscosity grade, the time required for the polymer powder to dissolve ranged from hours to days on a shaker. Trapped bubbles were removed from the solution using a vortex mixer. The yellowish viscous solution was transferred to high-density polyethylene (HDPE) bottle and stored in the refrigerator until it was used. The solution was allowed to reach room temperature before dispensing.

Gel Bead Synthesis and Continuous Washing

Ca-Fe gel beads were prepared as described in previously published works (Min and Hering 1998 a, b, US Patent No. 6,203,709), and an overview of the process is summarized here.

Calcium alginate beads were prepared by delivering the sodium alginate aqueous solution drop-wise into a continuously stirred calcium chloride solution. This technique allows the formation of spherical droplets, which, after being dispersed in $CaCl_2$ solution, result in formation of spherical beads. A peristaltic pump was used to dispense the polymer solution. The tip at the end of the dispensing tube was connected with the pipette tips to make beads with different sizes. A dispensing tip was positioned approximately 7 cm above the surface of 0.1 M $CaCl_2$ solution used for gel formation. The clearance was needed between the tip of the dispenser and the surface of the medium (e.g., calcium solution), so that when a biopolymer droplet was dispensed, it would have sufficient time to change from teardrop shape as it left the tip to spherical shape as it entered calcium chloride solution.

The tubing and dispensing setup were flushed a few times with the polymer solution to be dispensed. The setup was then programmed to dispense a drop of polymer approximately every 3 seconds into 1 L of 0.1 M $CaCl_2$ solution in a beaker. When droplets were allowed to form freely at a pipette tip by dispensing the polymer solution slowly, they grew until the weight of the droplet exceeded the surface tension force at the tip. A propeller type impeller was used instead of a stir bar to provide gentle stirring from the surface of the solution, since the beads were reported to be susceptible to hydrodynamic forces (Ogbonna et al. 1991).

After dispensing the appropriate amount of the sodium alginate solution drop-wise, each beaker containing the fresh spherical beads was transferred to a paddle stirrer unit, and the solution with gel beads was stirred slowly (Figure 5.1). The acid-resistant polypropylene paddles were used as paddle assemblies for all mixing and washing because some of the cation and washing solutions were acidic.

The beads were allowed to cure in $CaCl_2$ solution for three days and then removed from the solution and rinsed twice with Milli-Q water to make sure that there were no unbound cations on the surface of the beads. The beads were then transferred to a continuous washing system (15 mL/min) for 3 days. For preparation of Ca-Fe beads, washed calcium beads were transferred into 1 L of 0.1 M $FeCl_3$ solution (Figure 5.2), which allowed partial exchange of calcium for Fe(III). The beads were continuously washed for 3 days with 1 mM HCl (15 mL/min) as shown in Figure 5.3 and stored in 1 mM HCl until used.

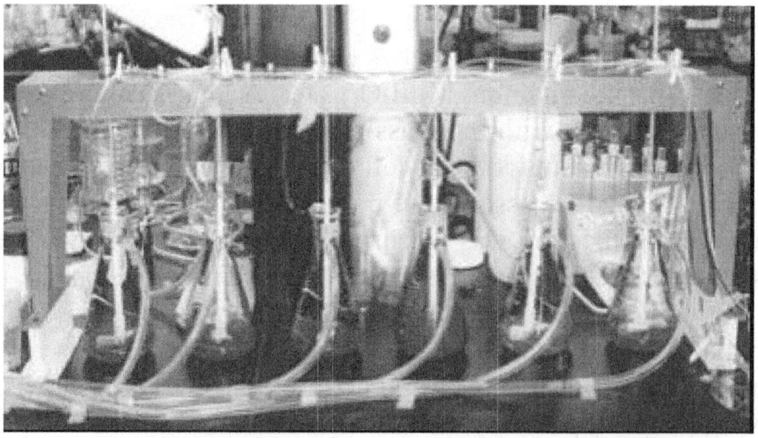

Figure 5.1 Calcium-alginate gel beads curing in 0.1 M $CaCl_2$ solution

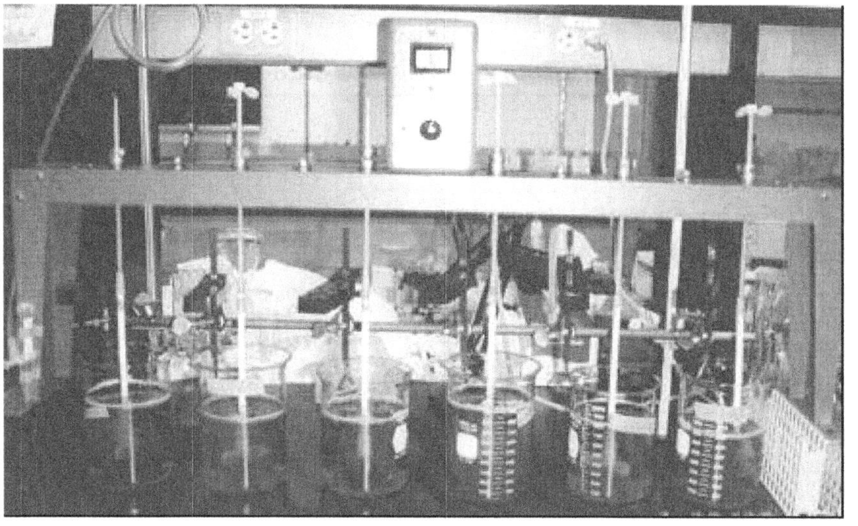

Figure 5.2 Doping process for calcium-alginate gel beads in 0.1 M $FeCl_3$ solution

Figure 5.3 Washing step for Ca-Fe gel beads with 1 mM HCI solution

RESULTS AND DISCUSSION

Stability of the Ca-Fe Gel Beads

The stability of Ca-Fe gel beads was determined in different test solutions as summarized in Table 5.1. In some solutions with high TDS, the Ca-Fe gel beads were found to lose their physical integrity; the stability of the beads under varying conditions was assessed based on the time at which dissolution of the beads was observed. This effect was not simply a function of the ionic strength of the solutions but was strongly dependent on pH and also on the specific ions present. The disintegration of the gel at higher pH values is most probably due the dissociation of carboxyl acid residues, which results in collapse of the gel structure (Yotsuyanagi et al. 1990), and allows migration of water into the gel. Electrostatic repulsion among charged residues in the beads is also a possible explanation for the swelling and dissolution (Yotsuyanagi et al. 1991).

Table 5.1
Composition of test solutions and bead stability

Composition of test solutions			Time before bead dissolution	
Constituent	Concentration (mg/L)	pH of solution	At original pH of test solution	At pH = 4
$NaCl$ $NaHCO_3$ $NaNO_3$ Na_2SO_4	20,651 9,912 1,071 8,378	8.01	2 hours	12 days
KCl K_2SO_4 KNO_3 $NaHCO_3$	26,299 10,266 1,273 9,912	7.35	2 hours	12 days
$MgSO_4$ $Mg(NO_3)_2$ $MgCl_2$ $NaHCO_3$	7,080 933 3,306 9,912	8.08	5 days	8 days
$MgCl_2$ $Ca(NO_3)_2$ K_2SO_4 $NaHCO_3$	17,100 1,033 16,815 9,912	7.55	5 days	8 days
Na_2SO_4 $NaCl$ $NaNO_3$ $NaHCO_3$ $CaCl_2$ K_2SO_4 $CuCl_2$ $FeCl_3$ $CoNO_3$ $MgSO_4$ $MnSO_4$ $NaHPO_4$	4,544 64,467 4,335 4,116 965 8 1.3 1.7 0.1 13 0.04 0.03	7.94	1 hour	8 days

The stability of Ca-Fe gel beads could be maintained for over a week by lowering the pH to 4. Elevated carbonate concentrations were found to be a major factor in destabilizing the beads. This provides a simple method of digesting the beads (for analytical purposes) but will require decarbonation of the brines for treatment purposes. The pH of 4 provides both maximum stability to the beads and, as shown in previous studies (Min and Hering 1998a), optimum performance in arsenic uptake.

Table 5.2
Iron content of Ca-Fe gel beads

Sodium alginate			
Type	Manufacturer	Concentration % (w/v)	Iron amount (μmol)
High M	Sigma Co.	2	0.552
High M	Sigma Co.	4	0.818
High G	Wako Chemicals	2	0.641
High G	Wako Chemicals	4	0.822

Iron Content Analysis

Iron content of the beads was determined by spectrophotometry using phenanthroline (Table 5.2). For measurements, beads were dissolved in bicarbonate solution and then filtered using 0.2 μm cellulose acetate filters (Whatman). Iron calibration standard solutions were prepared with the same amount of sodium alginate by disintegrating the appropriate number of calcium alginate beads that were collected before doping with $FeCl_3$ solutions. These standards were used to verify that there was no significant interference of these for the iron analysis by the analytical method used (Clesceri et al. 1989).

Batch Studies of Arsenic Uptake from Brines by Ca-Fe Gel Beads

Arsenic uptake from brines was tested with four brine solutions (Table 5.3). The simple brine was designed to represent the TDS of the La Puente ion exchange brine samples (drum A) as described in Chapter 6 of this report except that perchlorate was omitted and the nitrate concentration was increased to compensate for the reported carbonate concentration. For the sulfate, phosphate, and sulfate/phosphate brines, the simple brine was amended to include representative concentrations of these anions, which could be expected to compete with arsenate for sorption sites on the Ca-Fe gel beads. Nitrate concentrations were correspondingly decreased to minimize changes in ionic strength (TDS).

Table 5.3
Composition of brines tested for arsenic uptake

	Concentration of constituents (mg/L)						
Brine type	NaCl	$NaNO_3$	Na_2SO_4	$CaCl_2$	KCl	$MgCl_2$	Na_2HPO_4
Simple	47,180	21,845	-	59.8	6.3	9.2	-
SO_4	47,210	17,128	3,955	59.8	6.3	9.2	-
PO_4	47,210	21,845	-	59.8	6.3	9.2	7.2
SO_4/PO_4	47,210	17,128	3,955	59.8	6.3	9.2	7.2

For the batch experiments, 50 mL polycarbonate centrifuge tubes were used as reactors with 25 or 30 ml of solution as reaction volume. All of the batch experiments were done at pH 4. Beads were equilibrated using a pH 4 solution for a day and transferred to the reaction tubes after the surface moisture of the beads was wiped off. Then brine solution was added and the reaction tubes were placed in a rack, which was mounted on a wrist action shaker for mixing. After the specified reaction times, the beads were removed from the tubes and the supernatant and gel beads were analyzed for arsenic.

Performance of Ca-Fe Gel Beads Synthesized from "High M" Sodium Alginate

The kinetics of arsenate uptake was examined for each of the brine compositions. Ca-Fe gel beads used for the kinetics study and further experiments described in this section were prepared using 2% "High M" sodium alginate. Residual arsenic concentrations were measured in the brine solutions; solutions were diluted with 1% nitric acid by a factor of 100 for initial arsenic concentrations up to 20 mg/L and by a factor of 500 for higher concentrations before analysis by ICP-MS, and calibration curves were prepared in a comparable (diluted) brine matrix. Arsenic accumulated in the Ca-Fe gel beads was also measured by first dissolving the beads in 10,000 mg/L bicarbonate solution, filtering through 0.2 μm cellulose acetate membrane filters (Whatman), and then diluting the filtrates by a factor of 100 with 1% nitric acid. Calibration standards were prepared in the same matrix to verify that matrix effects did not interfere with ICP-MS analysis. The mass balance of dissolved and sorbed arsenic for all samples ($n = 50$) was $99 \pm 13\%$. Over 90% of the sorption observed over 96 hours occurred in the first 48 hours of the time course experiments. For further batch experiments, an equilibration time of 120 hours was chosen to ensure complete reaction.

Removal of arsenate from solutions equilibrated with Ca-Fe gel beads was observed for all brine compositions (Figure 5.4). The sorption density (i.e., ratio of sorbed As(V) to Fe content) increased with increasing dissolved As(V) concentration and was observed to reach a plateau at elevated dissolved As(V) concentrations (Figure 5.5). Note that here, and elsewhere, "dissolved arsenic concentration" refers to the residual dissolved arsenic concentration after equilibrium with the Ca-Fe gel beads. For the phosphate and sulfate/phosphate brines, As(V) uptake was examined only in the lower range of initial As(V) concentrations; in this concentration range, similar sorption densities were observed for all brine compositions (Figure 5.6).

$$\{As\}_{sorb} = \frac{K\,[As]_{diss}\,\{AlgFe\}_T}{1 + K\,[As]_{diss}} \qquad (5.1)$$

Where $\{As\}_{sorb}$ is the sorption density [g As/g Fe], $[As]_{diss}$ is the dissolved (equilibrium) As(V) concentration [mg/L], $\{AlgFe\}_T$ is the maximum sorption capacity [g As/g Fe], and K is the conditional affinity constant [L/mg]. The values of the fitting parameters K and $\{AlgFe\}_T$ for the simple and sulfate brines are listed in Table 5.4. The values of K are not significantly different for the two brines but the maximum sorption density for As(V) is depressed by about 14% in the presence of sulfate.

Uptake of arsenate from brines by the Ca-Fe gel beads could be modeled using Equation 5.1.

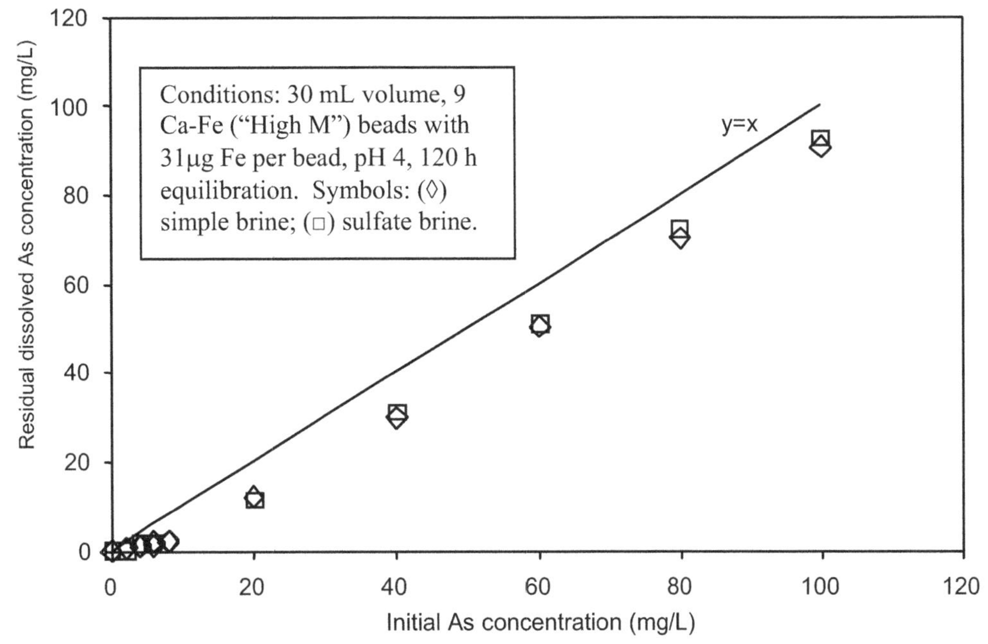

Figure 5.4 Residual dissolved arsenate concentrations as a function of initial (total) arsenate concentration after equilibration with Ca-Fe beads

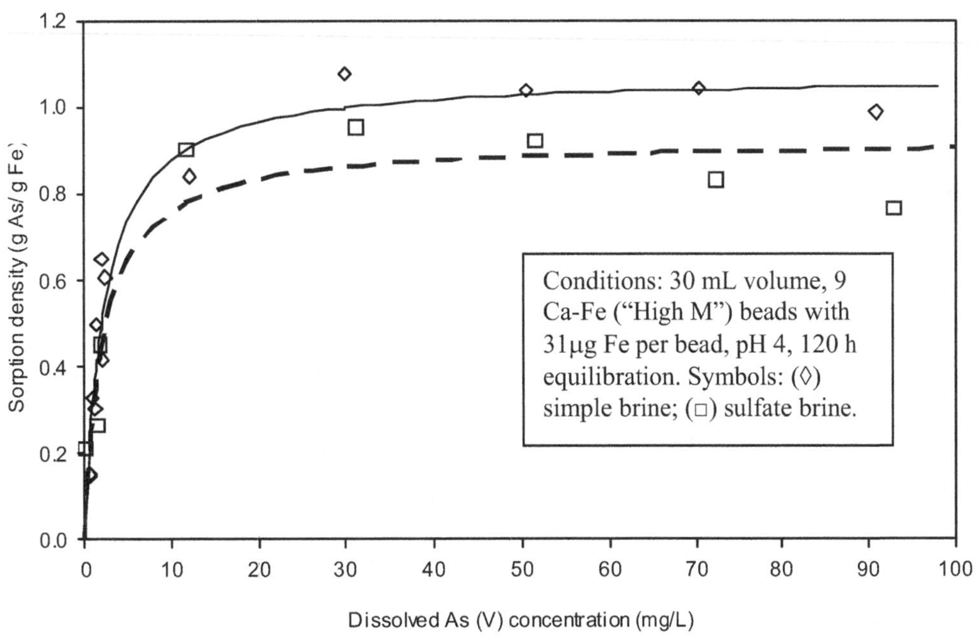

Figure 5.5 Sorption density (g As/g Fe) as a function of dissolved (equilibrium) As(V) concentration (lines show model fits)

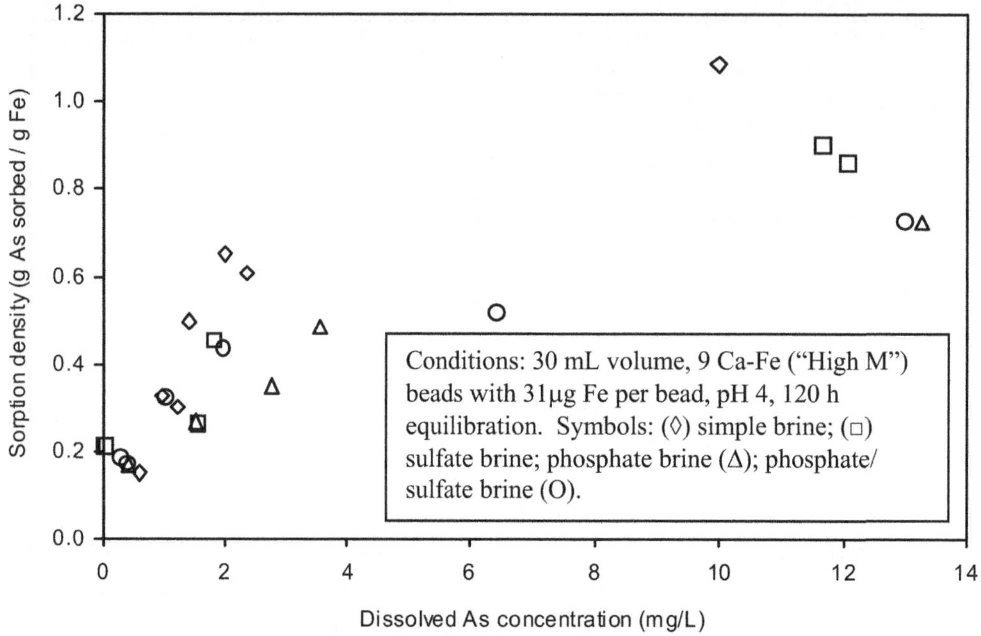

Figure 5.6 Sorption density (g As/g Fe) as a function of dissolved (equilibrium) As(V) concentration

Table 5.4
Fitting parameters for model of arsenic uptake by Ca-Fe gel beads synthesized from "High M" sodium alginate

	Type of brine	
Fitting parameters	Simple	Sulfate
K	0.442	0.463
{AlgFe}T	1.074	0.923

The efficiency of As(V) removal from brine solutions could be improved by increasing the iron loading of the system (effectively the number of Ca-Fe gel beads) as shown in Figure 5.7. The increase in As(V) removal as a function of iron content was consistent with the sorption behavior discussed above as indicated by the predicted removal shown as lines in Figure 5.7.

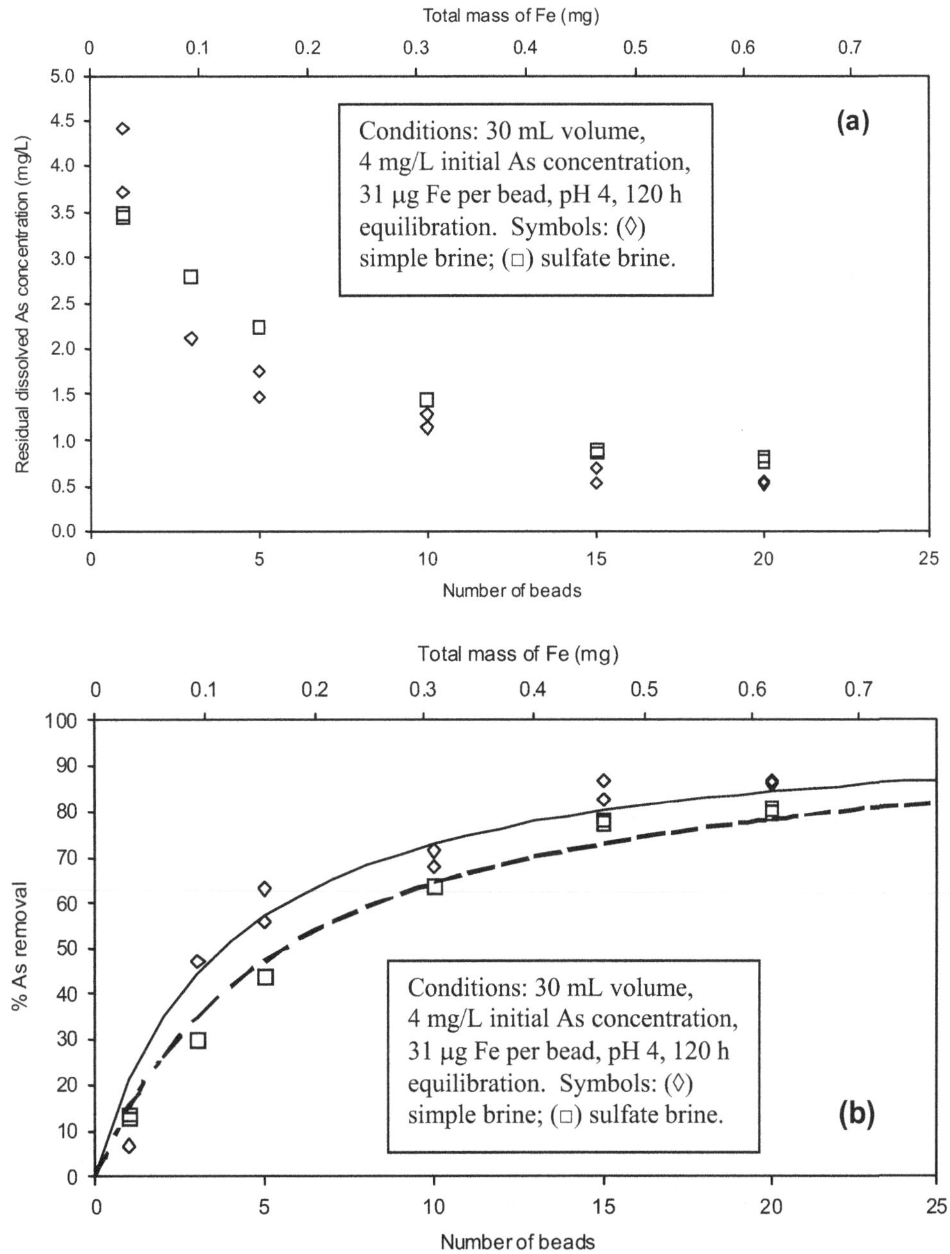

Figure 5.7 Efficiency of As(V) removal as a function of Fe content or corresponding number of ("High M") Ca-Fe beads; (a) residual dissolved As(V) concentrations, (b) % As(V) removal (lines show model fits)

Performance of Ca-Fe Beads Synthesized from "High G" Sodium Alginate

In order to investigate the effect of sodium alginate type, Ca-Fe beads prepared from 4% "High G" sodium alginate was used to remove As(V) from all brine solutions. The sorption density increased with increasing dissolved As(V) concentration (Figure 5.8 and Figure 5.9). The efficiency of As(V) removal from brine solutions could be improved by increasing the number of Ca-Fe beads (Figures 5.10 and 5.11).

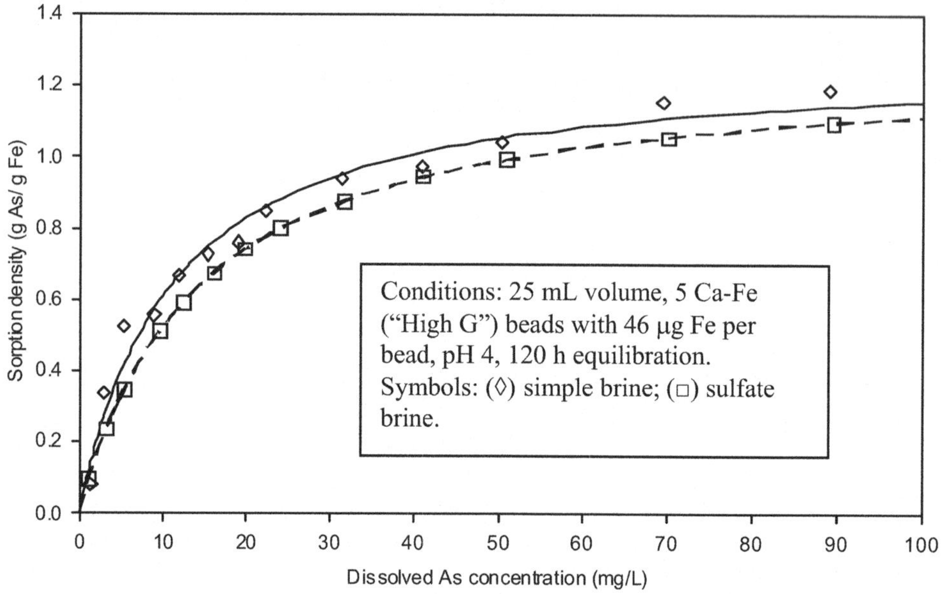

Figure 5.8 Sorption density (g As/g Fe) as a function of dissolved (equilibrium) As(V) concentration (lines show model fits)

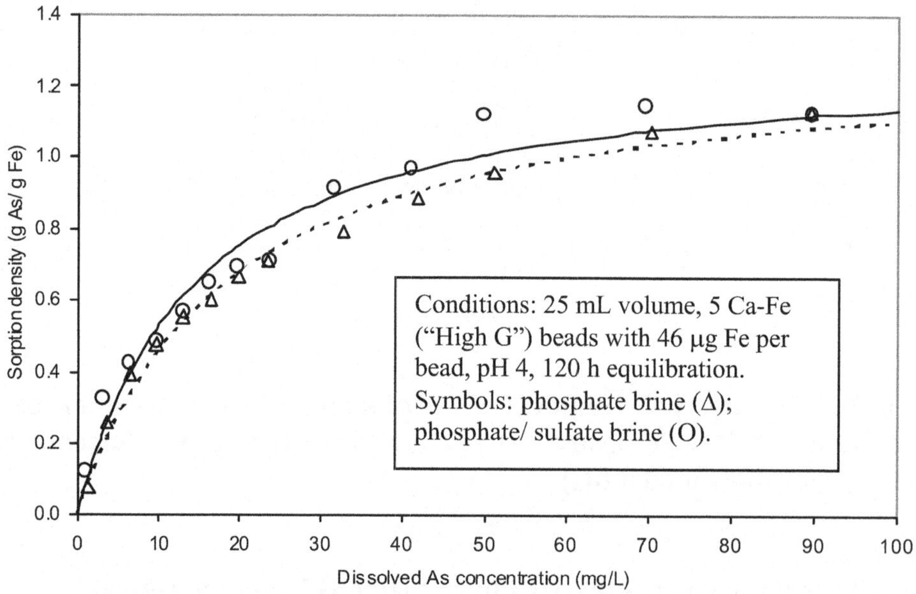

Figure 5.9 Sorption density (g As/g Fe) as a function of dissolved (equilibrium) As(V) concentration (lines show model fits)

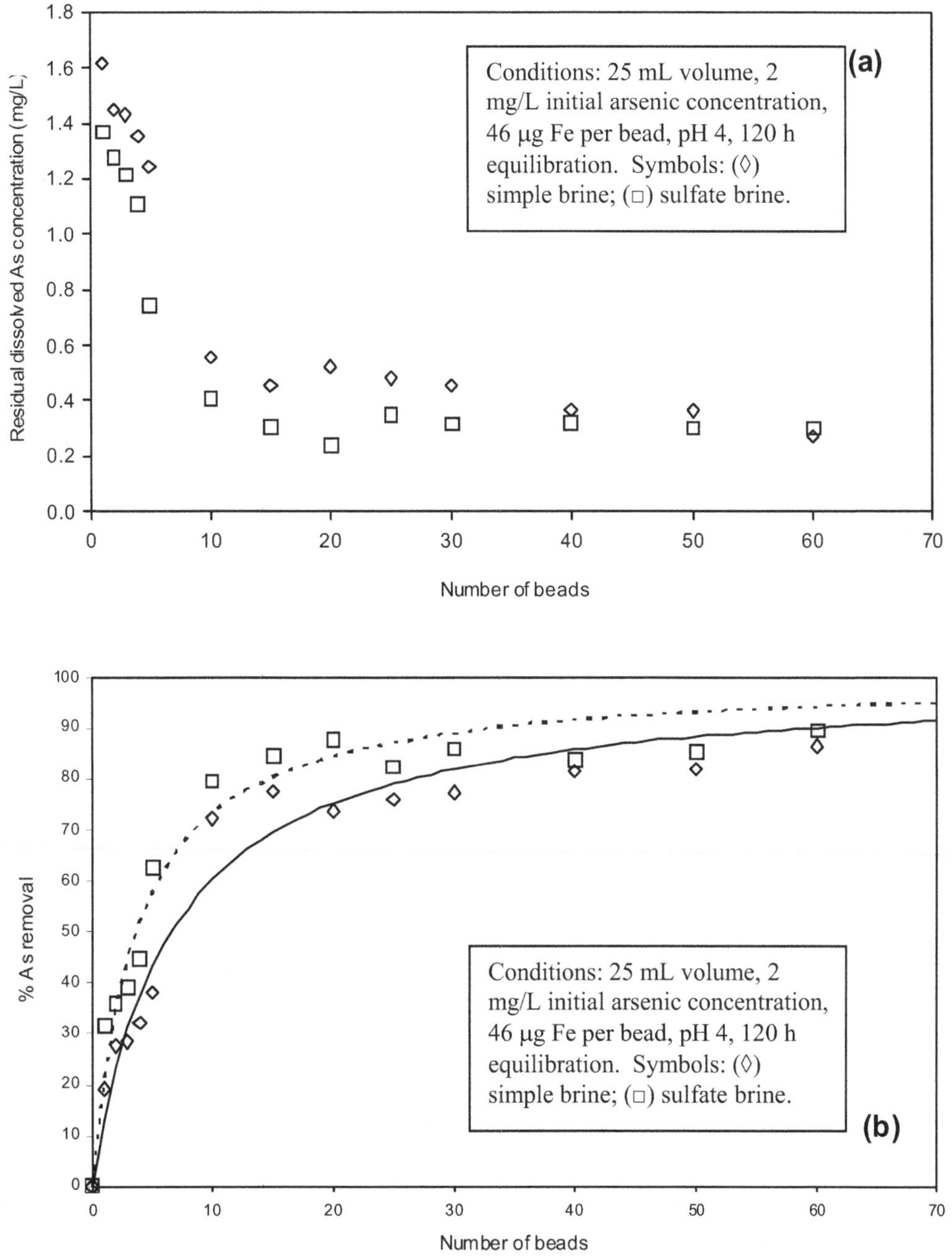

Figure 5.10 Efficiency of As(V) removal as a function of Fe content or corresponding number of ("High G") Ca-Fe beads for simple and sulfate brines, (a) residual dissolved As(V) concentrations, (b) % As(V) removal (lines show model fits)

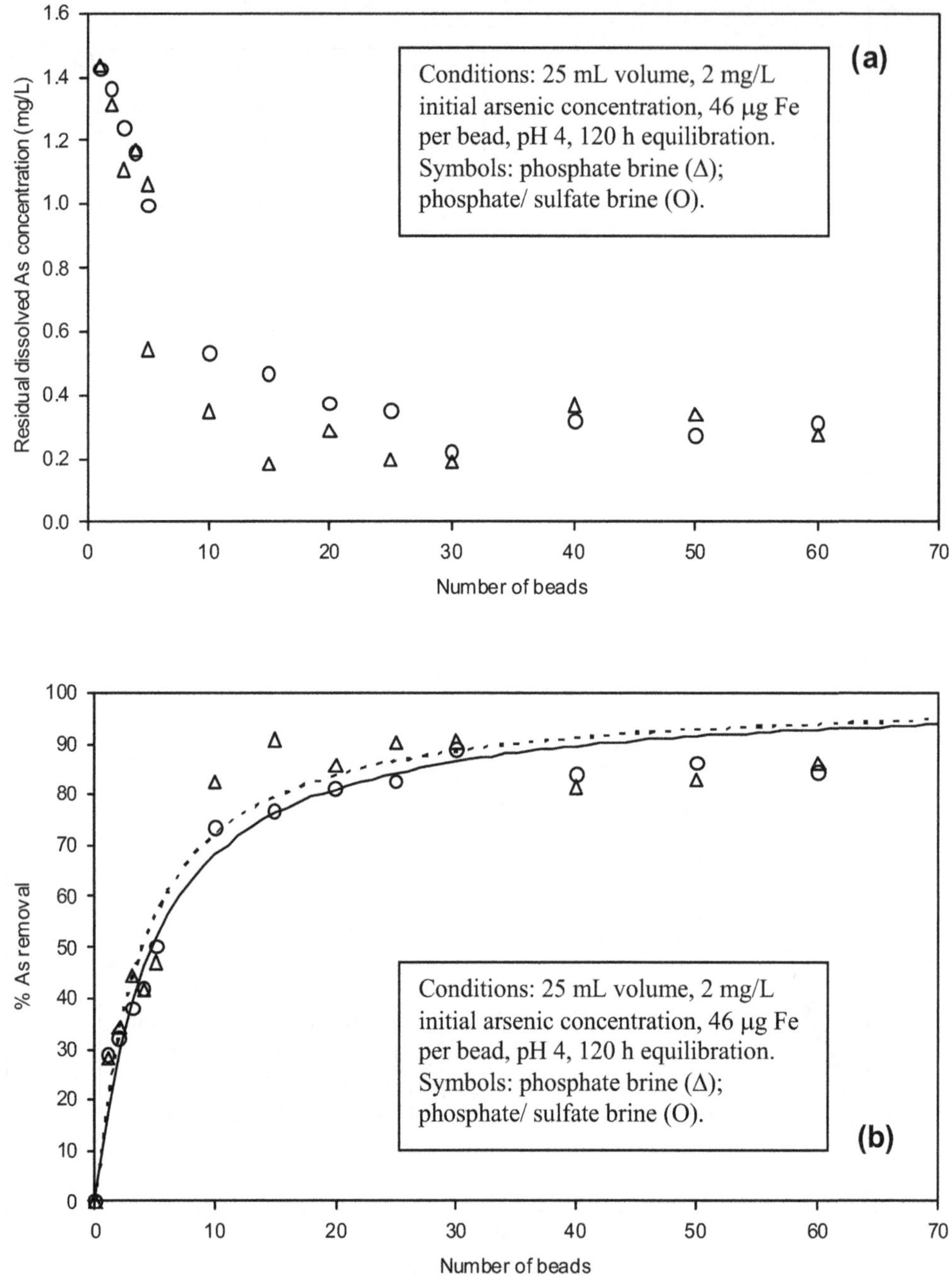

Figure 5.11 Efficiency of As(V) removal as a function of Fe content or corresponding number of ("High G") Ca-Fe beads for phosphate and phosphate/sulfate brines, (a) residual dissolved As(V) concentrations, (b) % As(V) removal (lines show model fits)

Table 5.5
Fitting parameters for model of arsenic uptake by Ca-Fe gel beads synthesized from "High G" sodium alginate

Fitting parameters	Type of brine			
	Simple	Sulfate	Phosphate	Phosphate/sulfate
K	0.092	0.071	0.069	0.054
{AlgFe}T	1.282	1.270	1.298	1.317

Values of fitting parameters K and $\{AlgFe\}_T$ for model of As(V) uptake by Ca-Fe beads synthesized from "High G" sodium alginate for all brines are listed in Table 5.5. The values for K (which correspond to the strength of the association of As(V) with the Ca-Fe beads) listed in Table 5.5 for the simple and sulfate brines are smaller than the corresponding K values for the "High M" Ca-Fe gel beads (Table 5.4). This difference in affinity of the two types of Ca-Fe gel beads for arsenic is evident in Figures 5.5 and 5.8, where the higher K value for the "High G" beads corresponds to the steeper slope of the curve in Figure 5.5.

Some difference in the sorption behavior of the "High G" and "High M" beads was expected since cation (i.e., Ca^{2+} or Fe^{3+}) binding is attributed primarily to interactions with G blocks (Yotsuyanagi et al. 1990). However, it is possible that, for the formation of ternary complexes (i.e., alginate-Fe(III)-As(V)), on excess G blocks is undesirable. "High G'' Ca-Fe beads had higher sorption capacity than the Ca-Fe beads obtained from "High M" sodium alginate because the amount of iron in Ca-Fe beads synthesized from 4% "High G" sodium alginate is higher than the 2% "High M" Ca-Fe beads (Table 5.2).

Preliminary Investigation of HFO-Loaded Gel Beads

In addition to the substitution of Fe(III) for Ca in the Ca-Fe beads, an alternative strategy for incorporating Fe(III) into the alginate gel matrix was examined. In this alternative preparation, Fe(III) was incorporated into the gel as HFO rather than by the substitution of Fe(III) for Ca. Comparison of the performance of the two sorbents, the Ca-Fe ("High M") gel beads and HFO-loaded gel beads, could provide insight into the mechanism(s) underlying the performance of the Ca-Fe gel beads as well as indicating whether the HFO-loaded gel beads warrant further study.

HFO-loaded gel beads were prepared by dispersion of 2 g of sodium alginate ("High M") into 100 mL of 0.05 M HFO suspension at pH 5 and the suspension was dispensed drop-wise into a continuously- stirred calcium chloride solution. The beads were allowed to cure in $CaCl_2$ solution for 3 days and then removed from the solution and rinsed twice with Milli-Q water to make sure that there were no unbound cations on the surface of the beads. The beads were then continuously washed for 3 days with Milli-Q water (15 mL/min) and freshly used.

The efficiency of arsenate removal from the various brines was examined as a function of iron content for the HFO-loaded gel beads (Figure 5.12). Surprisingly, the efficiency of As(V) removal shows less improvement with increasing Fe content for the HFO-loaded beads than was observed for the Ca-Fe gel beads. Removal efficiencies were similar for the various brine compositions (Figure 5.13). The maximum observed removal efficiency was only 70%, corresponding to a residual dissolved As(V) concentration of 1.04 mg/L. These results did not justify continued investigation of the HFO-loaded gel sorbent.

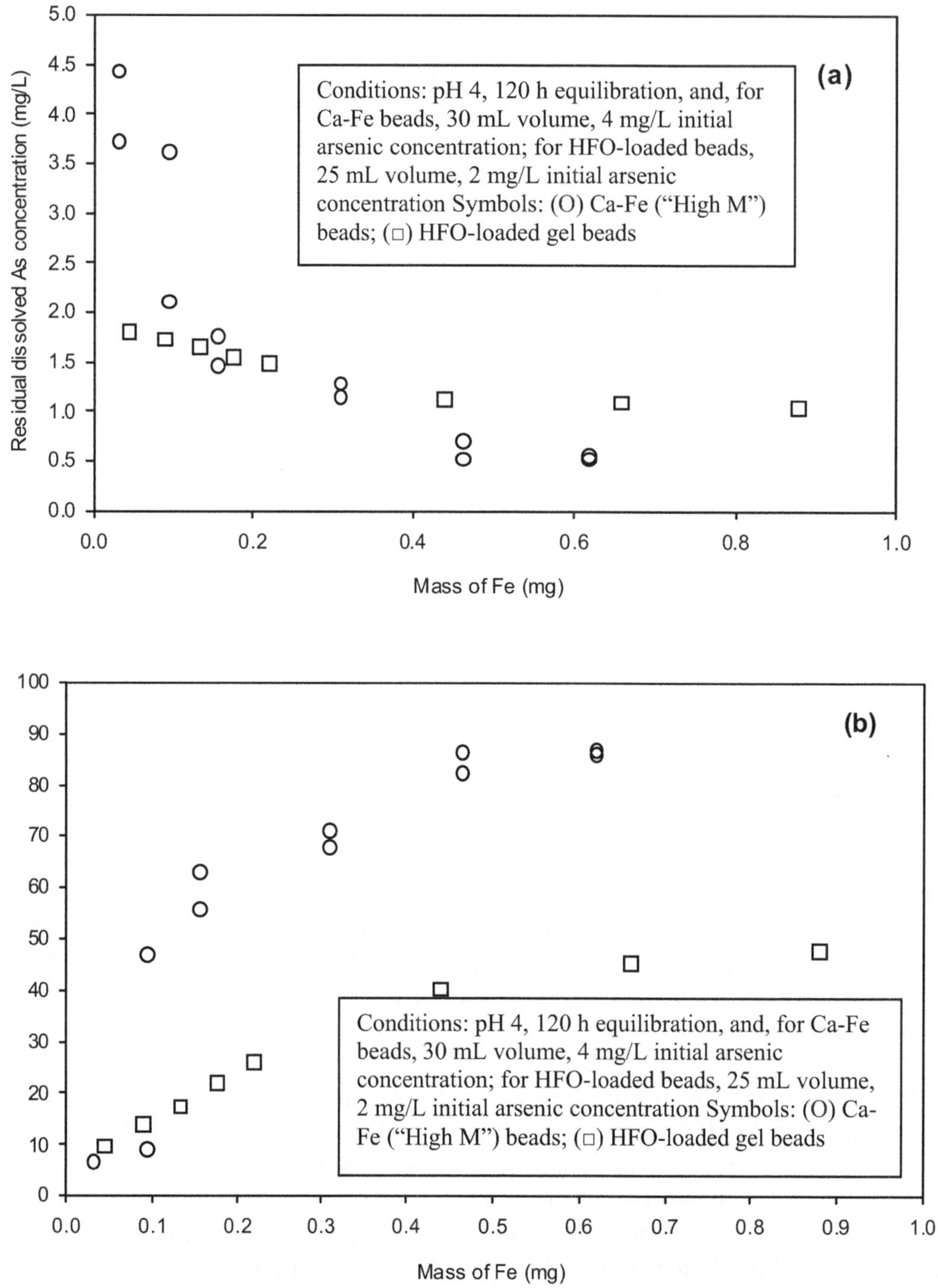

Figure 5.12 Efficiency of As(V) removal from simple brine as a function of Fe content for Ca-Fe and HFO-loaded gel beads, (a) residual dissolved As(V) concentrations, (b) % As(V) removal

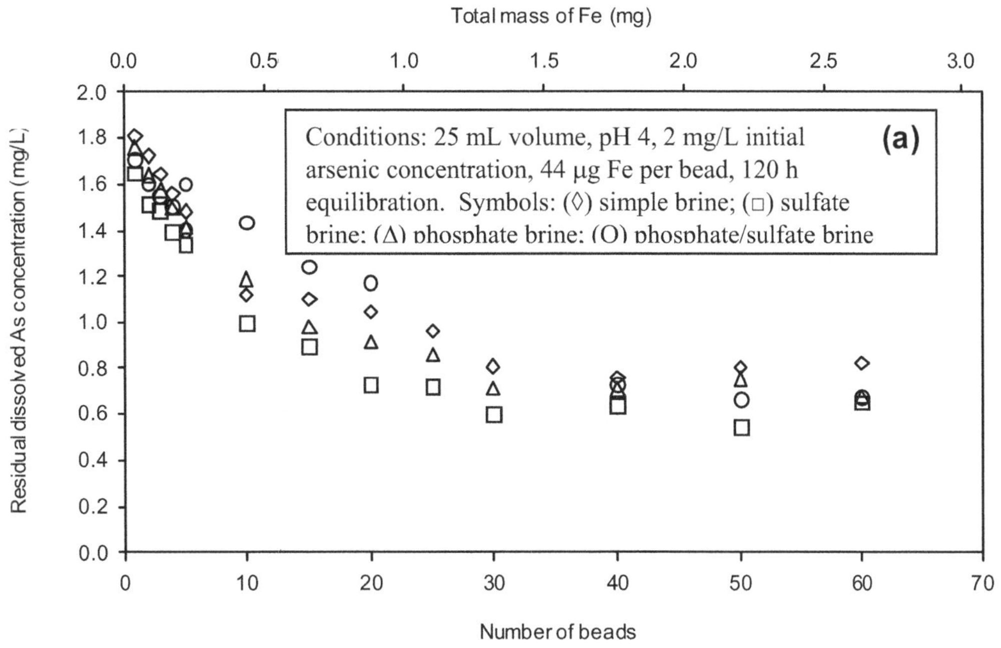

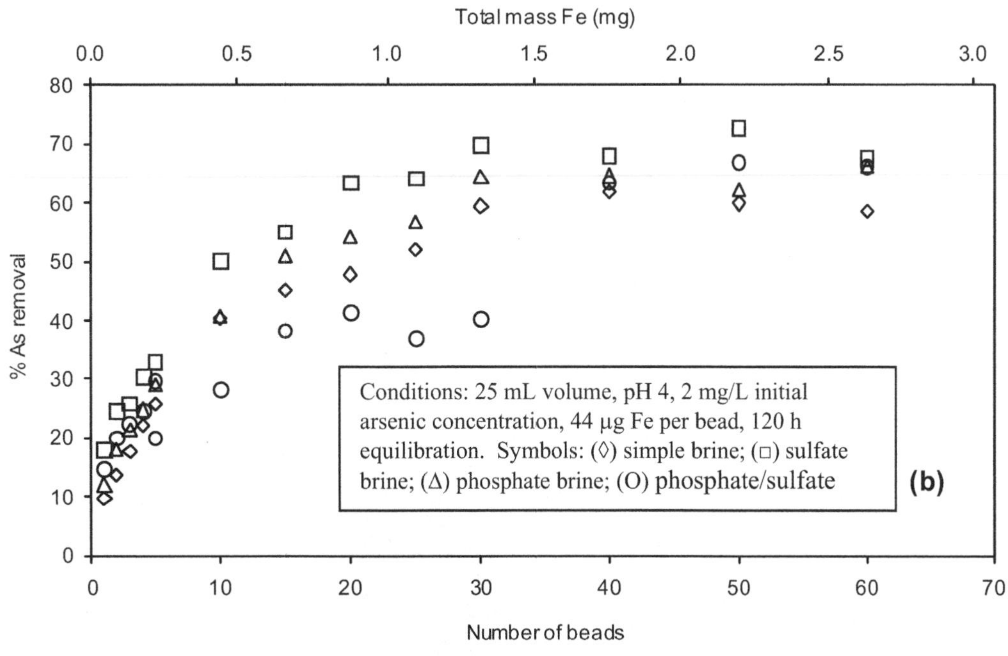

Figure 5.13 Efficiency of As(V) removal by HFO-loaded gel beads as a function of Fe content for various brines, (a) residual dissolved As(V) concentrations, (b) % As(V) removal

Leaching Tests on Dehydrated Spent Ca-Fe ("High G") Beads

Initial tests were conducted with spent beads dried at 65°C in an oven until constant mass was obtained. After dehydration, the size of beads was reduced to approximately 3.6% of the volume and 5.3% of the mass of original beads (Figure 5.14).

For the TCLP, the dried spent beads were extracted with 0.2 M sodium acetate buffer (pH = 5) for 18 hours with a weight of solution equal to 20 times that of the beads and, for the WET, with 0.2 M sodium citrate buffer (pH=5) for 48 hours with a weight of solution equal to 10 times that of the beads.

In the WET test after the 48-hour extraction time, the spent beads lost their physical integrity and the entire extract became somewhat gelatinous. The solution and gel phases could not be cleanly separated either by centrifugation or filtration. It was not possible to remove the solution by pipette without also transferring some of the gel phase. Thus, it appears that the Ca-Fe gel beads are not compatible with the conditions of the WET test without additional stabilization.

Under the TCLP conditions, less than 1% of the arsenic in the spent beads was extracted (Table 5.6). However because of the small volume of extraction solution used, the arsenic concentration in some extracts exceeded 1 mg/L. Therefore, drying conditions were varied to see if performance in the TCLP test could be improved.

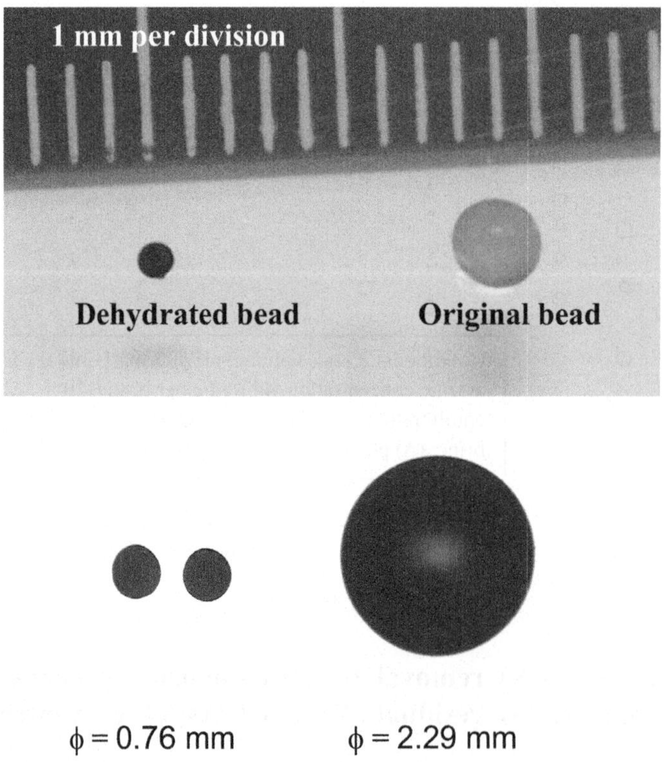

Figure 5. 14 Dehydrated and hydrated forms of Ca-Fe beads along with their typical sizes

Table 5.6
Performance of dehydrated spent beads in TCLP test

Brine type	Mass of As in spent beads (mg)	Mass of As extracted from spent beads (mg)	Percent As extracted from spent beads	As in extract (mg/L)
Simple	0.20	0.0017	0.86	1.31
SO4	0.14	0.0077	0.56	0.59
PO4	0.12	0.0056	0.48	0.49
SO4/PO4	0.29	0.0049	0.17	1.46

Effect of Drying Temperature on the Performance of Spent Beads in TCLP Test

The effects of temperature on calcium alginate gels have been reported previously (Serp et al. 2002). In this study, calcium alginate gel beads were exposed to temperatures ranging from 25°C to 130°C for varying periods of time resulting in expulsion of water, substantial shrinkage, and increase in the alginate density. The heat-treated gels were found to be less permeable and more resistant to citrate and other complexing agents than gels dehydrated at lower temperatures.

In order to investigate the effect of drying temperature on the performance of spent Ca-Fe gel beads in TCLP tests, the temperatures of 24, 65, 94, and 127°C were selected as drying temperatures for Ca-Fe gel beads that had been used in the uptake of arsenic from simple brine.

In contrast to the results of the previous study, dehydration of spent Ca-Fe beads at increasing temperatures did not improve their performance in the TCLP test (Table 5.7). Rather, the mass of arsenic in the TCLP extract increased with increased drying temperature.

Table 5.7
Effect of drying temperature on the performance of spent beads in TCLP test

Drying temperature (°C)	Mass of As in spent beads before TCLP test (mg)	Mass of As extracted from spent beads (mg)	Percent As extracted from spent beads
24	1.758	0.005	0.3
65	1.758	0.015	0.8
94	1.758	0.11	6.4
127	1.758	0.13	7.2

Column Studies of Arsenic Uptake from Simple Brine Solution

Uptake of As(V) from simple brine solution by the Ca-Fe ("High G") beads was investigated in column studies. Experiments were performed at pH 4. Before packing the column, the Ca-Fe gel beads were pre-equilibrated at pH 4 to allow for any swelling or shrinking. A 16 mL column with 1 cm inner diameter and 20 cm length was packed with 1,600 Ca-Fe beads of 2.3 mm diameter. The void volume of the packed column was 5 mL. The column was run in an up-flow mode by using a peristaltic pump with the flow rates of 12 and 6 mL/h. The influent As(V) concentration was 150 mg/L in simple brine. The calculated residence times of the column were 25 minutes for the flow rate of 12 mL/h and 50 minutes for 6 mL/h. Samples were collected every 50 minutes using a fraction collector and were diluted with 1% nitric acid before analysis by ICP-MS.

Despite the hundred-hour time scale required for equilibration in batch studies, efficient removal of As(V) from brine solution was accomplished in a column study with a residence time of approximately 25 minutes. The breakthrough curve shows that the ratio of effluent to influent As(V) concentration (C/C_0) reached 0.5 after 122 bed volumes (Figure 5.15).

Because of the high influent arsenic concentration used in this study ($C_0 = 150$ mg/L), the concentration of As(V) in the effluent prior to significant breakthrough ($C/C_0 = 0.05$ at 82 bed volumes) is not apparent in Figure 5.15. The initial breakthrough of As(V) is more clearly illustrated in Figure 5.16, which shows that the effluent arsenic concentration was < 10 μg/L up to 25 bed volumes and reached 1 mg/L at 64 bed volumes. It should be noted, however, that influent concentration used in this column test was much higher than what would be expected in ion exchange brines.

Good agreement was observed between the extent of arsenic removed in the column study and that expected in the basis of the batch experiments. Based on the maximum capacity determined from batch studies (1.282 g As(V)/g Fe) and the amount of Fe in the Ca-Fe gel beads packed in the column, the predicted capacity of the column would be 0.095 g As(V). With an influent concentration of 150 mg/L As(V) and a bed volume of 5 mL, the observed column capacity was 0.090 g As(V).

At the end of the experiment, the column was disassembled and the sorbed As(V) in the beads from three regions of the column were measured. The arsenic to iron molar ratio measured for these beads are presented in Table 5.8. The arsenic to iron molar ratio was only slightly higher at the bottom of the column (i.e., inlet) than at the top of the column (outlet), indicating good distribution of arsenic sorption throughout the column.

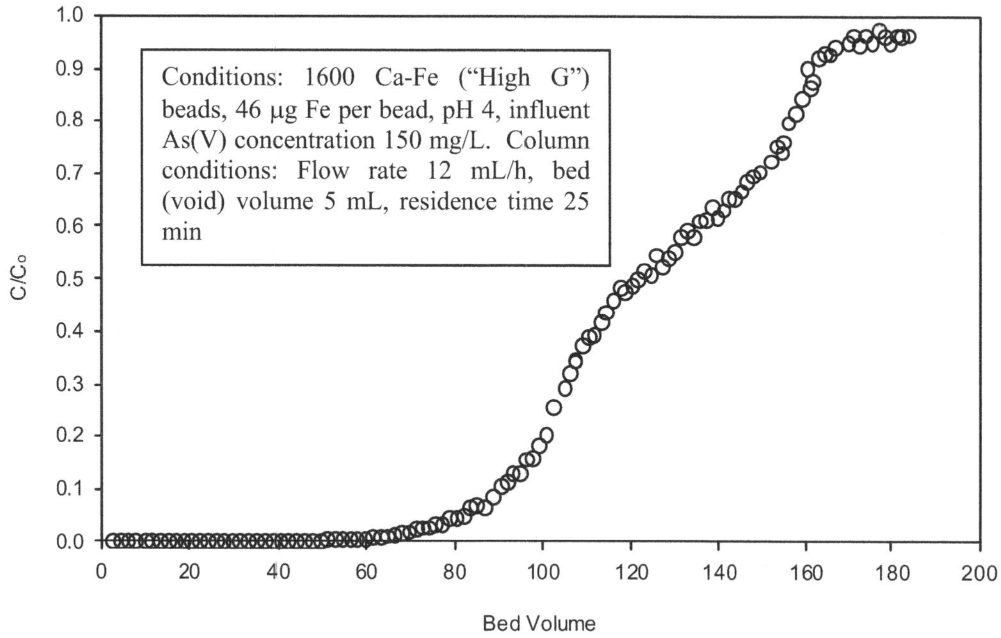

Figure 5.15 Breakthrough curve for removal of As(V) from simple brine using a column packed with Ca-Fe gel beads

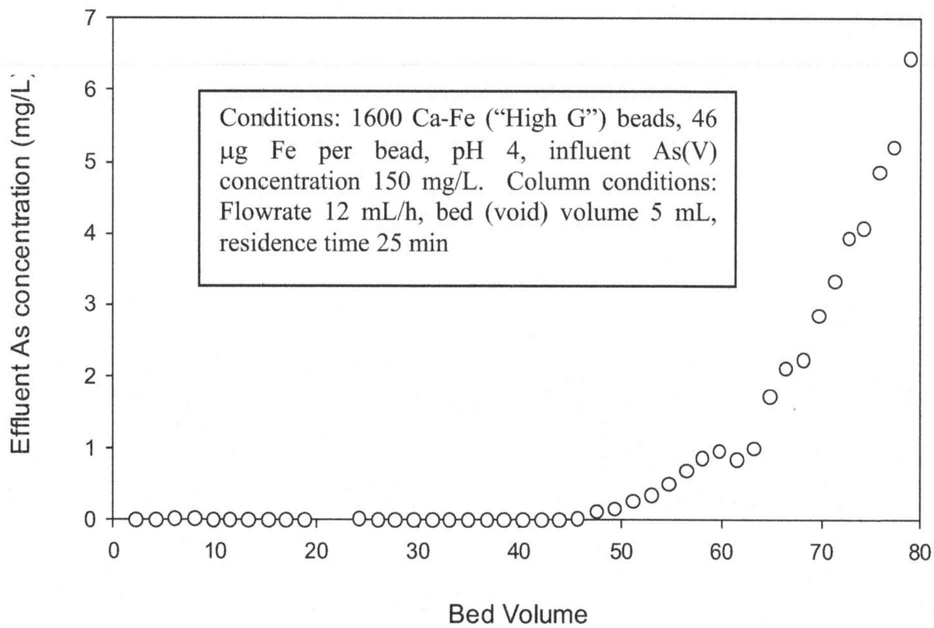

Figure 5.16 Breakthrough curve for removal of As(V) from simple brine using a column packed with Ca-Fe gel beads

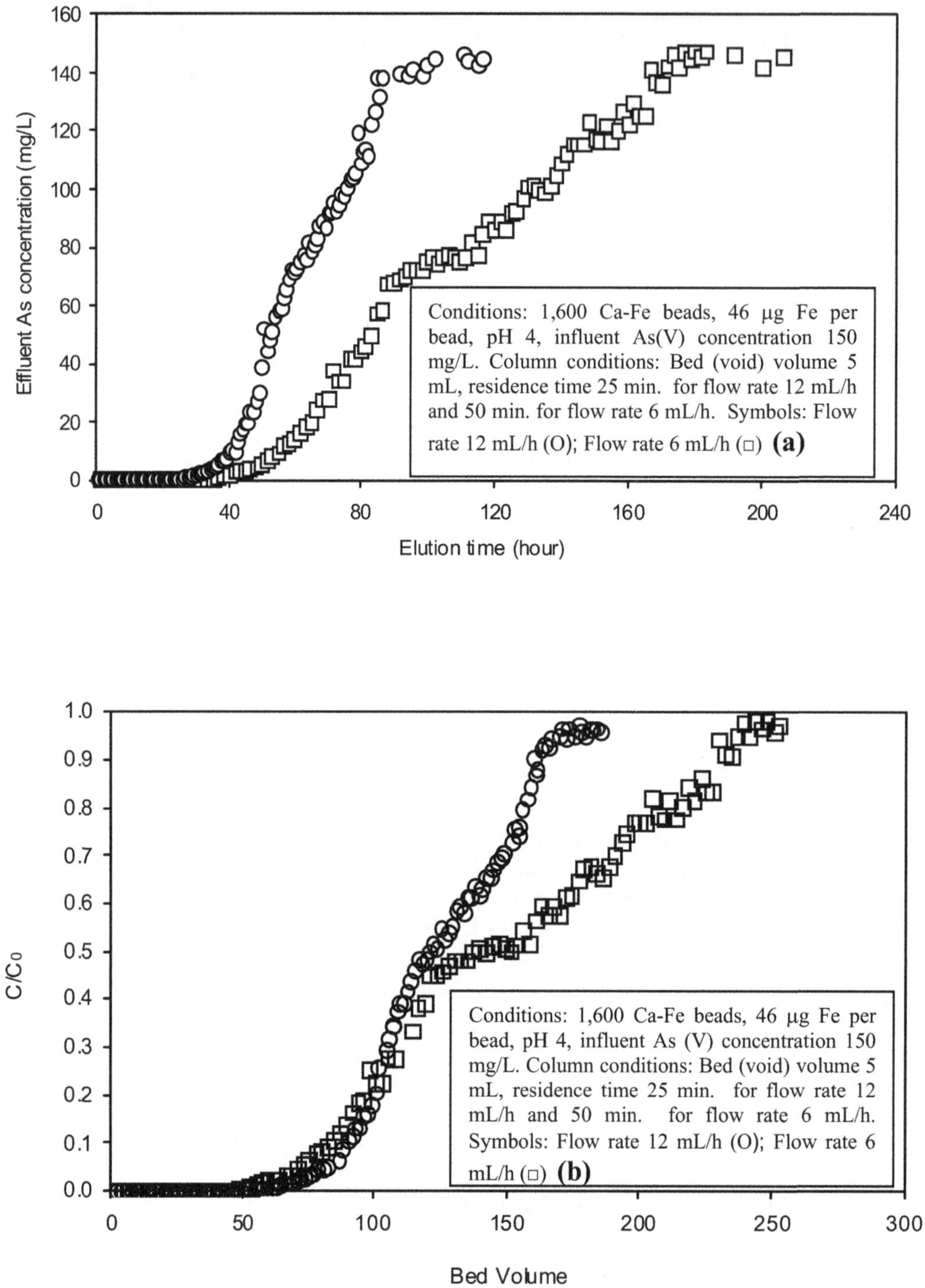

Figure 5.17 Breakthrough curve for removal of As(V) from simple brine using a column packed with Ca-Fe ("High G") beads, (a) effluent As(V) concentration, (b) ratio of effluent to influent As(V) concentration

Table 5.8
Mole of As(V) sorbed per mole of Fe from different location of the column

Location	As/Fe molar ratio
Top	0.88
Middle	0.89
Bottom	0.91

In order to test the effect of flow rate on column performance, a second column was prepared as described above and run under identical conditions except for the flow rate, which was decreased to 6 mL/h. The time to 50% breakthrough was corresponding increased from 50 hours to 83 hours (Figure 5.17a). When the ratio of effluent to influent As(V) concentration is plotted against bed volume (Figure 17b), C/C_0 reached 0.5 after 138 bed volumes for the flow rate of 6 mL/h.

SUMMARY AND RECOMMENDATIONS

Removal of arsenic from potable water by ion exchange or reverse osmosis generates arsenic-contaminated brines as a residual. Arsenic must be removed from the brines before disposal to evaporation ponds or to coastal waters through a brine line. Here, the application of an environmentally-benign, biopolymeric sorbent for the removal of As(V) from brines was examined, considering the effect of brine composition on As(V) uptake and the stability of the spent, dehydrated sorbent in the TCLP and California WET. The biosorbent was prepared as Ca-Fe gel beads, which were used in batch and column tests of As(V) uptake. Spent beads were dehydrated before leaching tests.

In the Ca-Fe gel beads, Fe is immobilized throughout the entire volume of the gel by coordination with carboxylate functional groups of the alginate polymer. Thus uptake of As(V) occurs throughout the volume of the gel bead rather than only at the surface. This allows As(V) sorption densities to reach very high levels (nearly 1:1 on a molar basis with Fe).

Unexpectedly, higher maximum As(V) sorption densities were observed in brines than had previously been observed in dilute salt solutions. As(V) sorption was depressed in the presence of phosphate and sulfate at levels representative of ion exchange brines, but As(V) uptake was still observed even with these competing sorbates in large excess of As(V). A significant limitation of the Ca-Fe gel beads, however, is their instability at neutral pH and in the presence of high bicarbonate concentrations. For this reason, all sorption tests were performed at pH 4.

The extent of As(V) removal in column tests was consistent with that observed in batch studies. Initial breakthrough in the column tests was unaffected by a two-fold change in flow rate. Later in the breakthrough curve the As(V) concentration in the effluent reached the influent value faster at the faster flow rate.

The spent, dehydrated beads were stable under the conditions of the TCLP test and released between 0.3 and 7% of their arsenic content, depending on the dehydration conditions. Under the conditions of the WET test, however, the dehydrated beads were not stable and formed a gelatinous mass from which the extract solution could not be cleanly separated. Disposal of the spent, dehydrated beads would therefore require additional stabilization.

The Ca-Fe gel beads hold promise for the treatment of As(V)-contaminated brines. In particular, the uptake of As(V) by the Ca-Fe gel beads is not inhibited by high salt concentrations as long as carbonate concentrations and pH conditions are controlled to maintain bead stability. These considerations would require lowering the pH to 4 and outgassing CO_2. Alternative methods of As(V) removal from brines would include coagulation (i.e., with ferric or aluminum salts) or adsorption onto fixed bed media (such as granular ferric hydroxide). Detailed information on the performance of these alternative removal processes in brines is not available for comparison.

However, the capacity of Ca-Fe gel beads (normalized to iron content) was greater than that generally observed for iron-based coagulants. As indicated previously, in coagulation, much of the iron is in the bulk material of the iron oxyhydroxide floc and is unavailable for arsenic binding. With the Ca-Fe gel beads, arsenic is adsorbed throughout the entire volume of the bead and a high proportion of the iron is available to bind arsenic.

In contrast to coagulant sludge, the spent beads are easily dehydrated, producing a material that is amenable to handling and transport. The spent, dehydrated beads are compatible with the TCLP test but are not stable under conditions of the WET test. Thus, in California, some additional stabilization of the spent, dehydrated beads would be required prior to disposal.

The alginate biopolymer used to prepare the Ca-Fe gel beads is a commercial product that is widely used in manufacture of low-cost food items. Preparation of the beads is a relatively simple process, which does not involve any specialty chemicals. Costs associated with large-scale production of Ca-Fe gel beads are therefore not expected to be prohibitive. However, further testing of the beads, at pilot-scale, is recommended to assess possible impacts of bead stability on performance over longer periods. The issue of stabilization for disposal in California requiring the WET test would also need to be addressed.

CHAPTER 6
PERCHLORATE AND NITRATE BIOLOGICAL BRINE TREATMENT

INTRODUCTION AND OBJECTIVES

One of the most widely adopted treatment processes for perchlorate and nitrate is ion exchange. During regeneration of ion exchange resin, perchlorate- and nitrate-laden sodium chloride brine is generated. The main objective of this task is to biologically treat sodium chloride brine containing perchlorate and nitrate prior to discharging the brine. Innovative aspects of this approach include biologically reducing perchlorate to non-detectable concentrations in brine with much higher concentrations of nitrate and total dissolved solids using a continuous flow reactor system. Detailed discussion and results on this task are presented in the following sections.

BENCH-SCALE TESTS MATERIALS AND METHODS

Surrogate Perchlorate and Nitrate Brine Composition

Prior to testing conducted on actual brine, a brief reactor test was conducted on a representative surrogate. The composition of the surrogate was based on the composition of the groundwater at La Puente Valley County Water District (LPVCWD), the predicted individual anion loading on the resin, and the assumption that the volume of the brine regenerant volume is approximately 1% of the treated groundwater volume. The surrogate formulation assumes that there is no dilution of the brine that is used for ion exchange resin regeneration and that the initial salt (NaCl) concentration was 7%. The surrogate was diluted to 50% with tap water to reduce the NaCl content to a level that would be biodegradable and representative of potential feed solutions. The surrogate formulation is provided in Table 6.1, and detailed results of the surrogate test are presented later in this chapter, Results of the Perchlorate Brine Treatment using Surrogate.

Table 6.1
Surrogate NaCl brine composition

Constituent	Surrogate (mg/L)	50% Surrogate (mg/L)
Cl^-	38,300	19,200
Na^+	28,300	14,100
SO_4^{2-}	3,060	1,530
HCO_3^-	3,000	1,500
NO_3^-	2,500	1,250
Ca^{2+}	60	30
Mg^{2+}	15	7
ClO_4^-	10	5
Conductivity	-	55.9 mS/cm
TDS	75,200	37,600

Perchlorate and Nitrate Ion Exchange Brine Source and Characterization

Perchlorate and Nitrate Brine Source

A 2,500-gpm, Calgon Carbon Corporation's ISEP ion exchange process has been in operation since spring 2001 in La Puente, California to remove perchlorate from drinking water for LPVCWD. The ISEP process is a pseudo-continuous, counter-current, ion exchange process, where ion exchange columns continually revolve on a rotating carousel and are either sequentially on-line, regenerated with sodium chloride brine, or rinsed with water. Each column is attached by a flexible line to a multi-port valve system that enables the proper sequencing of ion exchange columns. Figure 6.1 shows the ISEP process in operation in La Puente. The multi-port valve assembly can be seen above the ion exchange columns.

As the carousel rotates, the multi-port valve system periodically re-indexes to the new valve location. The new location determines the operational status of each ion exchange column. All system flows (groundwater, brine, and rinse water) are stopped briefly when the valves re-index.

Brine effluent from the ISEP process was collected by Carollo staff and shipped to ARA in Panama City, Florida on January 22, 2003. Six 15-gal drums of spent brine were collected from a port in the brine line that discharges to the brine sump located in the building that houses the ISEP system. The sample location is shown in Figure 6.2. This sample location is downstream of a manifold of three individual lines that transport spent brine and rinse water. It is also downstream of hydrochloric acid addition, which is used to prevent precipitation of calcium carbonate. Therefore, brine collected was diluted and acidified as discussed in the following section.

A detailed ISEP system schematic is shown in Figure 6.3. As all ion exchange columns slowly rotate on carousel-type support structure, a multi-port valve directs brine, rinse water, and groundwater flow to enable regeneration, rinse, and adsorption to occur simultaneously on different columns.

Figure 6.1 ISEP ion exchange System in La Puente, California

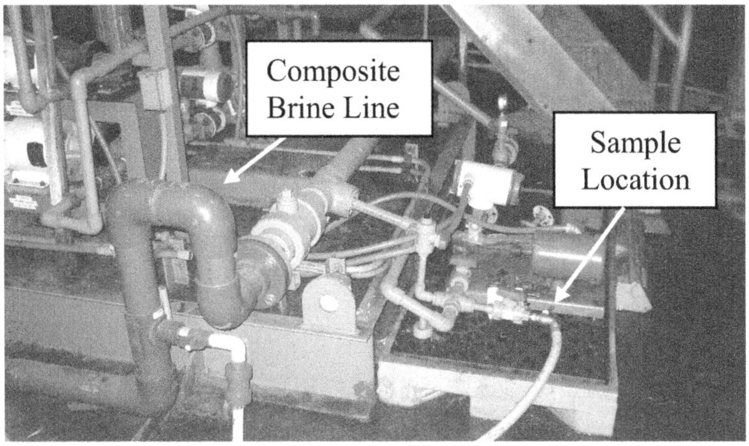

Figure 6.2 Perchlorate brine sample location at LPVCWD, California

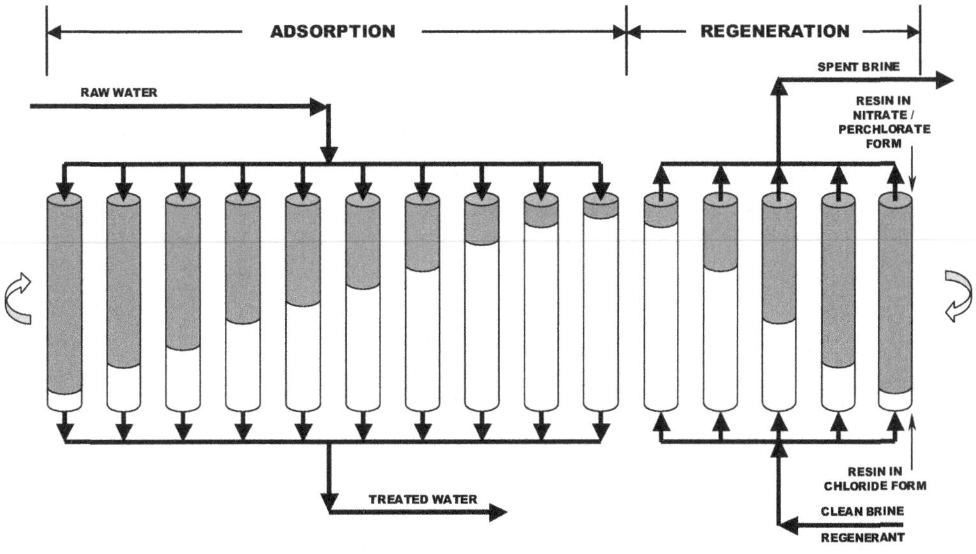

Figure 6.3 Calgon Carbon's ISEP continuous ion exchange system schematic (Source: Calgon Carbon Corporation 2002)

The valves of the ISEP system installed at LPVCWD are programmed to re-index every 23 minutes. At the time of re-indexing, different ion exchange columns come on-line and go off-line and there is a momentary break (~1 minute) in brine and rinse water flows. The six 15-gal samples were collected and labeled consecutively and sampling was initiated on the first drum immediately after the valves re-indexed. However, drum-sampling times did not directly coincide with valve indexing times. Therefore, as depicted in Figure 6.4, some samples were collected during valve indexing, which resulted in a sample from two different valve configurations at different times.

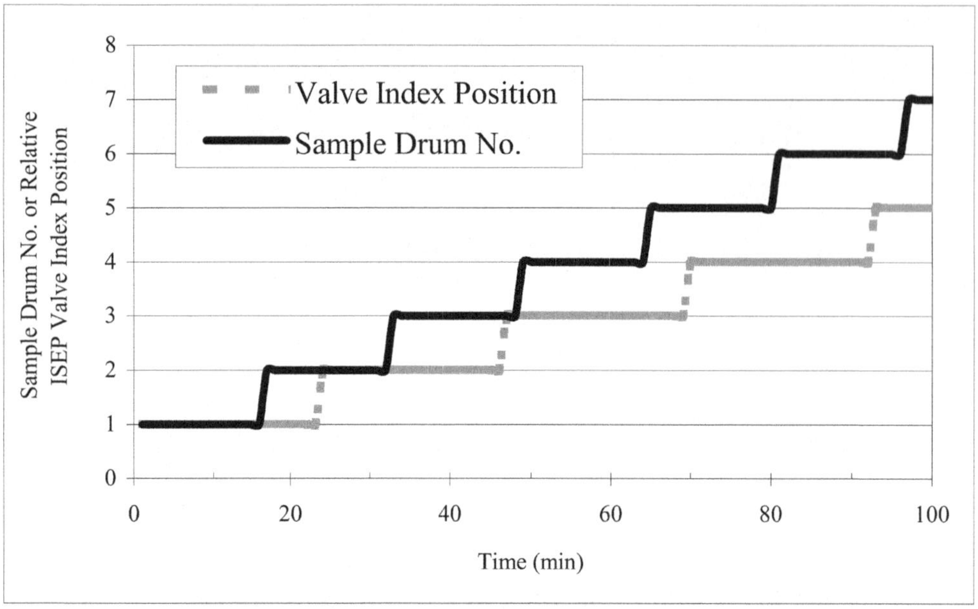

Figure 6.4 Brine sampling and ISEP valve indexing timing

The wide variation in brine properties observed for the first six samples raised questions concerning the variability of spent brine generated from the ISEP process. Three additional 15-gal samples were collected and pH was frequently measured. As can be seen in Figure 6.5, there was a broad variation in sample pH that did not exhibit a well-defined trend.

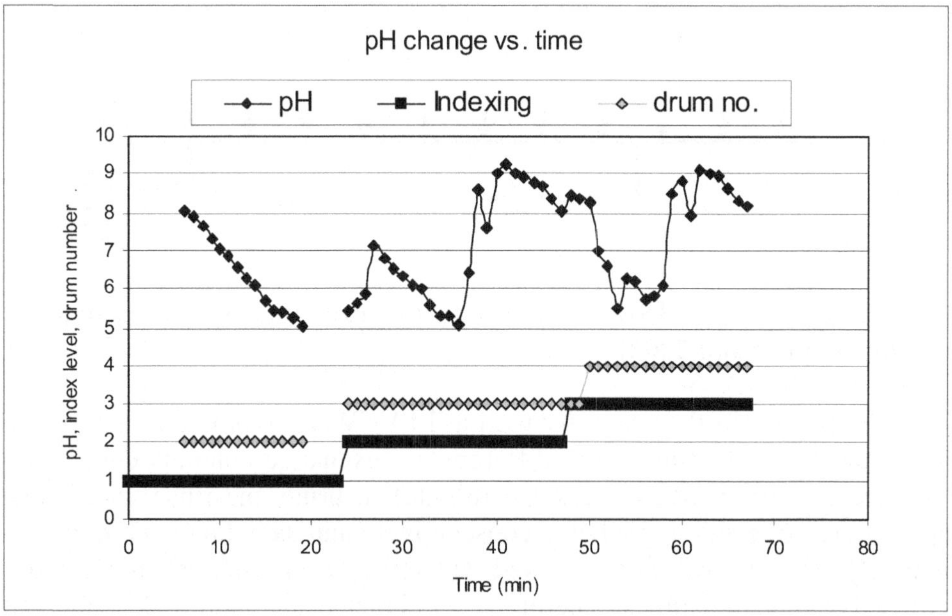

Figure 6.5 pH shift during valve indexing during additional brine sampling perchlorate/ nitrate ISEP brine characterization

ARA received the drums on January 30, 2003 and conducted initial analyses on the brine in each drum for perchlorate and conductivity. Two sets of three 15-gal drums were composited into two 55-gal drums. Selection of the 15-gal drums for each composite was determined from initial perchlorate and conductivity analyses shown in Table 6.1. It was assumed that these analyses would be representative of the composition and properties of the brine in each of the original 15-gal drums. Therefore, the contents of 15-gal drums labeled Nos. 1, 3, and 5 were transferred into composite drum "A," and the contents of 15-gal drums labeled Nos. 2, 4, and 6 were transferred into composite drum "B." A composite sample is more representative of an operational system that would include, by necessity, some feed equalization.

After composite drums A and B were prepared, additional analyses were conducted on samples retained from the original 15-gal drums. The results of the additional analyses are also provided in Table 6.2. As indicated, pH, sulfate, and nitrate concentrations varied greatly from drum to drum. Several factors may have contributed to the observed variability:

- Chromatographic elution of anions.
- The timing of sample collection.
- The continuous addition of hydrochloric acid during valve indexing.

After valves re-index, loaded column(s) begin brine regeneration. The brine selectively elutes anions from the loaded resin. This results in a chromatographic effect where individual anions will elute in slugs causing concentration peaks as a function of time. This effect, coupled with the timing of drum sampling, is the likely cause of the wide range of nitrate and sulfate concentrations observed. In addition, the continuous addition of the acid may have caused a pH decrease that was evident for drums that were collected during valve indexing.

Table 6.2
Analysis of individual La Puente ion exchange effluent brine samples

Analysis (mg/L)	Drum number (in order collected)					
	No. 1	No. 2	No. 3	No. 4	No. 5	No. 6
Initial analyses						
ClO_4^-	2.913	3.345	3.116	4.315	4.252	3.089
Cond. (mS/cm)	81.2	95.1	82.9	106.7	107.8	76.1
TDS (%)	5.90	7.09	6.04	8.17	8.28	5.48
Additional analyses						
NO_3^-	1,336	489	400	3,155	2,132	1,809
SO_4^{2-}	1,560	1,252	6,001	633	800	517
Cl^-	22,638	21,623	18,671	35,522	39,166	17,074
pH	7.25	0.89	7.18	6.83	4.91	1.24

Table 6.3 provides the anion analysis of the composite solutions contained in drums A and B. The nitrate concentration of each composite was similar. However, due to very high sulfate concentration of drum No. 3, sulfate concentration in composite drum A is over three times higher than that of composite drum B.

Table 6.4 provides solids and other analyses of the composite water quality. Solids analysis (total solids (TS), total suspended solids (TSS), and total dissolved solids (TDS)) by standard gravimetric methods was difficult to conduct because the high bicarbonate content of the brine made complete water evaporation difficult to achieve. However, there is a general agreement between the measured TDS and the TDS based on conductivity for sodium chloride solutions. The alkalinity of drum A was measured at approximately 8,000 mg/L. Since the pH of drum B was very low due to the low pH in the 15-gal sample drums 2 and 6, the alkalinity of drum B was not measured.

Table 6.5 provides the results of ICP metal analyses. Generally, the results agree with the expected values in the ion exchange brine. Sodium concentrations were consistent with chloride concentrations, and potassium concentration was lower than expected.

Table 6.3
Anion analysis of La Puente ion exchange brine composites (mg/L)

Drum	NO_3^-	NO_2^-	SO_4^{2-}	PO_4^{3-}	Cl^-	ClO_4^-
A	1,347	ND	2,849	ND	30,443	4.121
B	1,593	ND	900	ND	24,895	4.588

Table 6.4
Solids and other analyses of La Puente ion exchange brine composites (mg/L)

Drum	TS	TSS	TDS	TOC	TKN	pH	Cond. (mS/cm)
A	89,725	7.5	68,625	2.6	>0.20	7.17	91.6 (6.78% TDS)
B	53,500	5.0	51,980	2.2	0.21	1.61	80.5 (5.84% TDS)

TKN: Total Kjeldahl Nitrogen

Table 6.5
ICP analysis of La Puente brine composites (mg/L)

Drum	As	Ca	Co	Cu	Fe	K	Mg	Mn	Mo	Na	Ni	P
A	0.16	23	0.032	0.10	0.58	3.5	2.5	0.015	0.58	26,000	< 0.040	1.6
B	0.062	21	0.034	0.64	0.56	4.3	2.4	0.011	0.27	20,000	< 0.040	4.5

Summary of Analytical Methods

Analytical methods used for the brine sample analyses are summarized in Table 6.6. The same methods were used during the bench-scale testing throughout this project. Due to the high TDS of the brine, samples were diluted 10:1 before perchlorate analyses were performed. Dilution requirements for anion analyses vary, depending on the concentration of anions in the sample. Minimum reporting levels (MRL) for the critical anions are summarized in Table 6.7.

Table 6.6
Analytical methods used for biological brine treatment study

Parameter	Lab	Method	Method type
Perchlorate	ARA	EPA 314.0	Ion chromatograph
Anions*	ARA	EPA 300.1	Ion chromatograph
Metals	Certified lab	SW3010/6010	Acid digestion, ICP
Solids (TS/TDS/TSS)	ARA	SM 2540	Gravimetric
TOC	Certified lab	EPA 415.1	
TKN	Certified lab	EPA 351.2	
pH	ARA	SM 4500	Electrometric

* Anions measured include: NO_3^-, NO_2^-, Cl^-, SO_4^{2-}, PO_4^{3-}, and ClO_3^-

Table 6.7
Minimum reporting levels (MRL) for critical parameters (anions)

Analyte	Untreated brine MRL (mg/L)	Treated brine MRL (mg/L)
Cl^-	750	750
NO_2^-	100	10
ClO_3^-	150	15
NO_3^-	150	15
PO_4^{3-}	1,500	1,500
SO_4^{2-}	1,125	1,125
ClO_4^-	0.05	0.05

BENCH-SCALE TESTS RESULTS AND DISCUSSION

Surrogate Perchlorate and Nitrate Brine

Preliminary tests were carried out to demonstrate the performance of the ARA biodegradation process for the reduction of perchlorate and nitrate residuals in the surrogate media and to condition the biological reactors to be used with actual ion exchange brine.

Biological Reactor Bench Equipment Description

The reactor configuration used for these studies is shown in Figure 6.6. In the field demonstration process, brine feed dilution was accomplished by the addition of potable water directly to the reactor. However, in the bench studies the feed solution was pre-blended to the appropriate concentration and fed from a single feed tank to the first-stage reactor (R-1). A blend of carbon-based nutrient and micronutrients is also fed to the R-1 from a single feed tank. The carbon nutrient is a commercially available, low-cost ($30 to $40 per ton) desugared molasses byproduct of sugar beet processing. This nutrient, which is called condenser separator bottoms (CSB), is a viscous liquid that is storable and pumpable. Since CSB is an unrefined molasses product, it contains sufficient micronutrients for most applications. Table 6.8 lists some general properties of CSB. The ARA prototype at Thiokol has very successfully used CSB as the carbon source since May 1999. Phosphoric acid and ferric chloride solutions are premixed with the nutrient to provide additional phosphorous and iron respectively. Additional micronutrients are also premixed with the CSB nutrient. The pH is maintained in the reactors in the range of 7.0 to 7.5 by the addition of sodium hydroxide (NaOH).

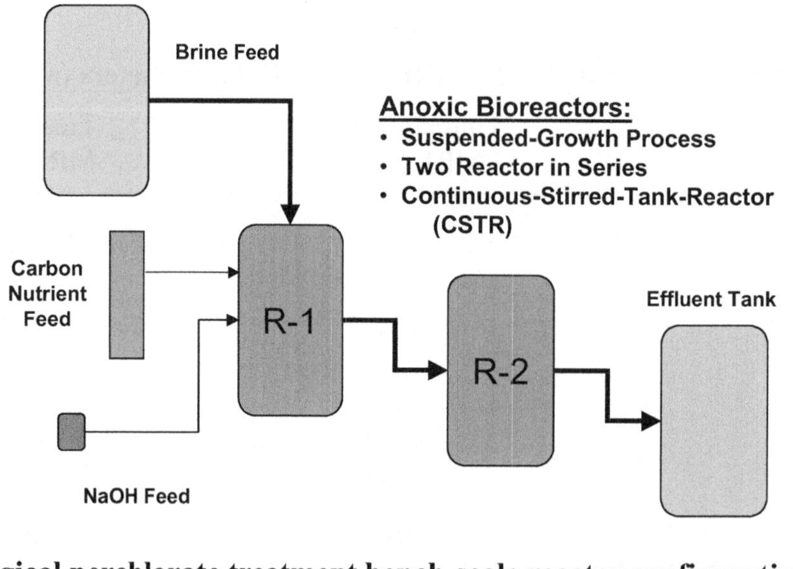

Figure 6.6 Biological perchlorate treatment bench-scale reactor configuration

Table 6.8
CSB properties

Analyses	Results
Dissolved Solids (%)	74
Ash (%)	22
Crude Protein (%)	16
Total Sugars (%)	17
TKN (%)	1.6
BOD (g/L)	312
pH	8-9

BOD: Biochemical Oxygen Demand

Figure 6.7 is a picture of the actual reactor system at ARA's Panama City, Florida facility. The reactors consist of two 3-L Applicon fermentation reactor systems. Each reactor has a working volume of approximately 2.5 L and is agitated. Temperature was controlled with an external water bath, which circulates heated water through internal heat exchanger coils. The process is effective over a broad temperature range (15 to 40°C), but it was controlled at 25°C in this study to eliminate temperature as a variable. Oxidation-reduction potential (ORP) was monitored to provide a general indicator of reactor performance. ORP is typically –300 to –400 in the R-1 and –400 to –500 in the R-2.

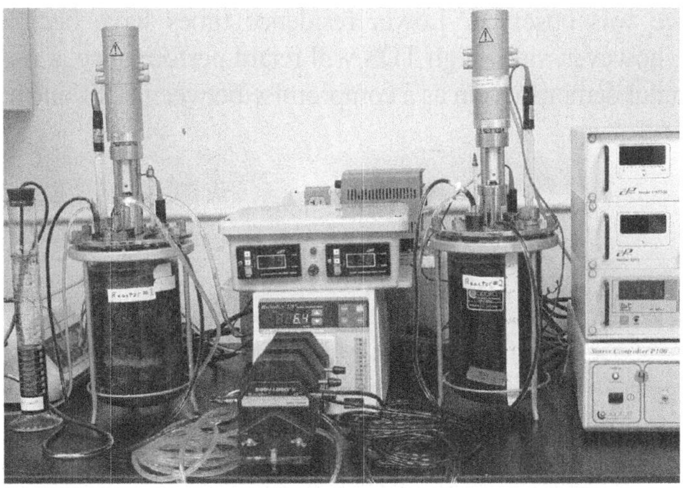

Figure 6.7 Bioreactor setup at ARA's Panama City facility for perchlorate treatment

Bench-Scale Reactor Inoculation and Startup on Surrogate

The R-1, was inoculated with mixed culture inoculums. All inoculums used for this study were derived from mixed cultures originally isolated in the early 1990s as part of projects conducted by ARA for the Air Force Research Laboratory. These cultures contain the *HAP-1/Wolinella succinogenes* organism and are maintained by ARA at their Panama City Research Facility.

Three different cultures that had been acclimated to three different media were used in the inoculation. The different media were: (1) high TDS; (2) high sulfate; and (3) low TDS, high perchlorate media. First, 2.25 L of 50% synthetic brine was added to R-1. Then 10 mL of CSB nutrient was added and the pH was adjusted to 7.5. Finally, 250 mL of mixed inoculums was added to the reactor. The reactor was allowed to operate in batch mode (no influent flow) overnight. Anion analysis the next day showed complete reduction of nitrate had occurred, which indicated a successful inoculation. Therefore, influent nutrient and brine flows were initiated.

Bench-Scale Biological Ion Exchange Brine Treatment Test Matrix

Critical operating variables of interest are the hydraulic residence time (HRT) and brine concentration because these two parameters directly affect process equipment size. Also, treating higher brine concentrations will result in reduced effluent flows and would facilitate brine recovery and reuse approaches. Nutrient and caustic consumption are important because they are the primary, non-labor components of operating cost. Typically, process start-up is accomplished by approaching target residence time and influent concentration incrementally, from less demanding conditions.

Table 6.9 is a summary of the surrogate test matrix that provides values of key operating variables for each test condition. The HRT is determined by dividing the water feed flow rate into the reactor working volume (2.5 liters). After the inoculation appeared successful, brine and nutrient flow was initiated at a rate, which would result in a 12-hour HRT. Typically, a 12-hour HRT is adequate for culture acclimation during startup. The HRT was gradually reduced to 8 hours, and then to 6 hours as acceptable performance was observed. Lower residence times have been demonstrated with this process (3 to 4 hours); however, since high TDS will retard performance, a minimum HRT of 6 hours was selected for this initial demonstration as a compromise between HRT and salt concentration.

Table 6.9
Biodegradation bench test matrix

Condition	Brine		HRT, hr	Nutrient ratio	Dates
	Source	TDS, %			
Inoculation	Surrogate	3.8	Batch	-	1/9/03
1 – Startup	Surrogate	3.8	12	1.2-1.88	1/10/03-1/23/03
2 – Reduce HRT	Surrogate	3.8	8	2.1	1/24/03-1/26/03
3 – Reduce HRT	Surrogate	3.8	6	2.1	1/27/03-1/30/03

The nutrient ratio is the mass ratio of available nutrient in the CSB to the mass of nitrate in the brine. This ratio is based on the organic carbon available in CSB, which is approximately 50% by weight. In a low TDS media that contains nitrate, a nutrient ratio slightly above 1.0 may be adequate. This ratio is partially based on the theoretical amount of alcohol that would be required for complete denitrification. Tchobanoglous and Burton (1991) derived an empirical equation for denitrification using methanol as the carbon source:

$$C_m = 2.47N_o + 1.53N_l + 0.87D_o$$

where: C_m = required methanol concentration, mg/L
N_o = initial nitrate-nitrogen concentration, mg/L
N_l = initial nitrite-nitrogen concentration, mg/L
D_o = initial DO concentration, mg/L

Ignoring nitrite and dissolved oxygen components of this equation, the methanol requirement equates to 0.56 mg methanol per 1.0 mg of nitrate. This does not account for carbon consumed in the reduction of other oxygenates that are present in spent ion exchange brine. Furthermore, it is anticipated that more nutrient is required in this high TDS environment in order to obtain acceptable denitrification and perchlorate reduction.

Results of the Perchlorate Brine Treatment Using Surrogate

The complete reduction of nitrate one day after inoculation indicated a successful inoculation and flow to R-1 was initiated. Typically, some nitrite formation is observed during start-up and transient conditions. Nitrite is an intermediate formed during nitrate reduction. Complete nitrate and nitrite reduction must occur before significant perchlorate reduction can be achieved. By the sixth day after inoculation (January 15, 2003), both nitrate and nitrite were being completely reduced in both reactors. Figure 6.8 shows that perchlorate reduction did not occur until January 18. The delay in perchlorate reduction may have been due to acclimation of the culture. On January 17, potassium phosphate (0.5 g/L K_2HPO_4) was added to the feed water to stimulate performance. Potassium was added because it is important in the regulation of osmotic pressure; therefore, increasing the potassium concentration may improve the general metabolism of the culture. A new batch of feed water was initiated on January 20 that did not contain potassium. Removing the potassium phosphate did not adversely affect performance; therefore, the onset of perchlorate reduction on January 17 was likely not the result of potassium addition. Reducing HRT on January 24 and January 27 did not adversely affect performance. The general trend observed is typical, and on average over 95% perchlorate was reduced in R-1. After January 20, the perchlorate concentration in second-stage reactor (R-2) was below the detection limit. A concentration of 0.01 mg/L was assigned for results that were below the detection limit. This concentration is approximately the detection limit in a clean matrix with a 10:1 dilution; however, the actual detection limit in this matrix could be 20 to 50 μg/L depending on interferences. Performance with the brine surrogate diluted to 50% (TDS of 37,600 mg/L) showed that complete nitrate and perchlorate reduction is readily achievable at the 6 hour HRT tested.

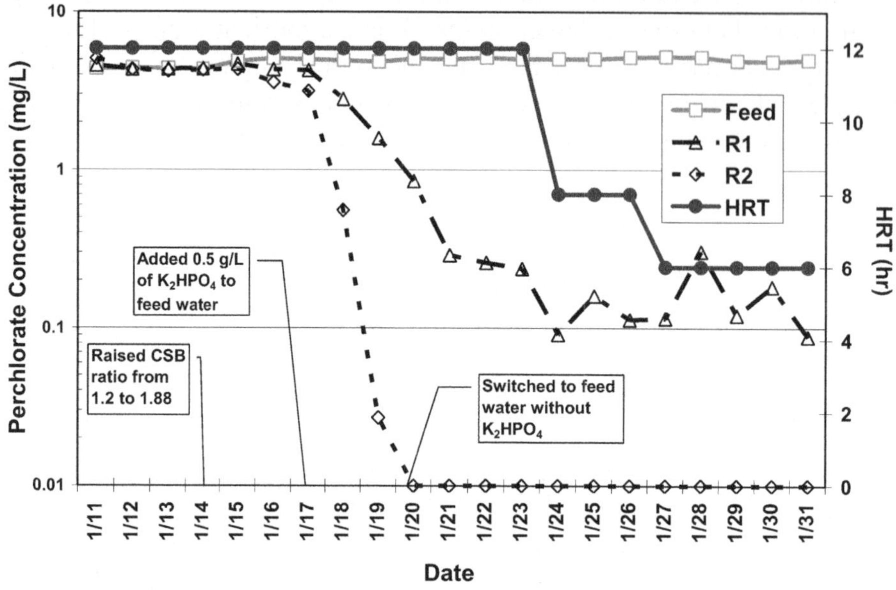

Figure 6.8 Bench-scale biological perchlorate reduction results with surrogate feed

Perchlorate, Nitrate, and Ferric Chloride Brines

A series of bench tests was conducted on several actual and surrogate brines. Table 6.10 provides a summary of bench testing conducted in the laboratory biodegradation reactor system previously described. Each test series evaluated performance for different types of feed brine. The bench-scale reactor system was cleaned and inoculated with fresh inoculums to begin this test series. Transition from one feed material to the next was accomplished without cleaning or re-inoculating the reactors. Changing feed matrix without re-inoculating eliminated inoculation startup and demonstrated that this culture can readily adapt to a broad range of environmental conditions. However, reactors were re-inoculated at the beginning of the 5% ferric chloride test series because a pH control failure resulted in a very low reactor pH that irreversibly damaged the microbial culture.

Table 6.10
Summary of bench-scale biodegradation testing

Brine feed tested			
Description	TDS (%)	HRT (hr)	Operating time (days)
ISEP NaCl brine from La Puente, California	1.74-2.94	8-12	41
ISEP brine with 1% surrogate $FeCl_3$	2.30-2.94	12-16	19
NaCl brine from Pomona, California	3.45	12-16	17
Ferric chloride surrogate brine	1.41-1.55	16	10
	Total Operating Time:		87

ISEP Brine Feed

Feed water for this test was brine from the Calgon ISEP process at the La Puente, California water treatment plant operated by the LPVCWD. Composite B (Tables 6.3, 6.4, and 6.5) was used for Test Conditions 1 through 7 as shown in Table 6.11. Reactor pH was controlled using 15% sodium hydroxide solution. Reactor temperature was controlled at 30°C to enhance performance. In general, increasing temperature up to 40°C increases microbial growth rate and performance. Thiokol maintains reactor temperature in the 30 to 35°C range when reducing perchlorate in high TDS media. The clean R-1 was inoculated with a combination of two inoculums: one adapted to high TDS water matrix and one adapted to high sulfate matrix. Brine feed was diluted to 30% concentration before inoculation. Approximately 21 hours after inoculation, flow was initiated to R-1 at an HRT of 12 hours.

Figure 6.9 shows the perchlorate concentration in the feed and each reactor (R-1 and R-2) as a function of days in operation. Minimal perchlorate reduction was observed in R-1 immediately after initiation of flow (Condition 1). However, after nutrient concentration was increased (Condition 2), complete perchlorate reduction was observed in R-2. As a result, reactor residence time was reduced to 8 hours. At 8-hour HRT, perchlorate reduction averaged approximately 80% in R-1 and 90 to 95% overall. Even after several days of operation, complete perchlorate reduction was not obtained due to an inadequate combination of residence time and nutrient concentration. When the residence time was increased to 12 hours (Condition 5), complete perchlorate reduction was obtained immediately in R-2. Results shown in Figure 6.9 as 20 μg/L was non-detect level, with a method detection limit of 20 to 40 μg/L. When brine feed concentration was subsequently increased to 40% (Condition 6), a one-day excursion in perchlorate concentration (51 μg/L) was observed in R-2. When brine concentration was increased to 50% (~30,000 mg/L total dissolved solids), complete perchlorate reduction was no longer maintained. Complete perchlorate reduction observed on the last day of testing was the result of an inadvertent reactor feed interruption.

Table 6.11
Test matrix for biodegradation of NaCl brine from the ISEP process

	Operating time (days)			Nutrient		Feed concentration		
#	Duration	Cumulative	HRT (hr)	Ratio	(g/L)*	Brine (%)	Conductivity (mS/cm)	TDS (%)
1	5	5	12	2.5	1.4	30	26.6	1.74
2	2	7	12	3.5	2.0	30	26.6	1.74
3	5	12	8	4.2	2.4	30	26.6	1.74
4	8	20	8	3.7	2.1	30	26.6	1.74
5	2	22	12	3.5	2.0	30	26.6	1.74
6	10	32	12	3.5	2.2	40	34.4	2.30
7	9	41	12	3.5	2.8	50	43.0	2.94

* Nutrient is based on organic carbon available in CSB (approximately 50% by weight)

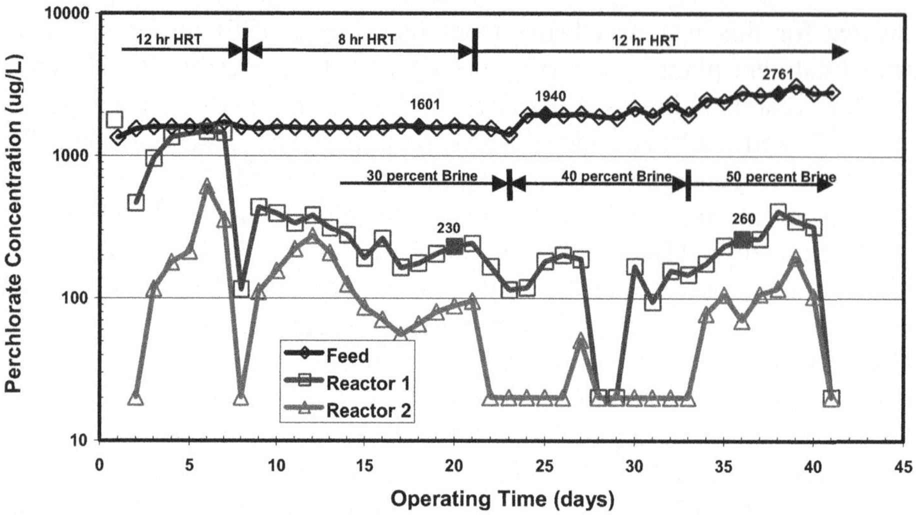

Figure 6.9 Perchlorate biodegradation performance - NaCl brine

During this test (Day 36), caustic for pH control was changed from sodium hydroxide to potassium hydroxide (KOH). The rationale for this change was to improve the metabolic functions of the bacteria by increasing the potassium concentration in a matrix with extremely high sodium and dissolved solids content. Changing the caustic did not have an apparent effect on performance. This test demonstrated that the proper combination of residence time and brine concentration permits complete perchlorate reduction.

On Day 32, the last day of testing at Condition 6 (40% brine), BOD, TKN, and solids analyses were conducted on effluent from R-2. The results of these analyses are provided in Table 6.12. The BOD and TKN result from the excess nutrient required to achieve complete nitrate and perchlorate reduction in this matrix. The measured BOD is consistent with predicted BOD based on the BOD of the nutrient and the nutrient concentration. Two factors contribute to the TKN observation. About 20% of the TKN is derived from organic nitrogen compounds in the nutrient (amino acids and proteins) and the remainder is from ammonia nitrogen generated under the strong reducing condition in the reactor. The volatile suspended solids (VSS) is a measure of the biomass that is generated by this suspended-growth process. Conventional denitrification processes generate 9 to 18% VSS based on nitrate concentration (Metcalf and Eddy, 1991). 135 mg/L equates to approximately 8.5%. The relatively low biomass generation is reflective of reduced growth rates in this high TDS media.

Table 6.12
Solids and other analyses

BOD (mg/L)	TKN (mg/L)	TSS (mg/L)	VSS (mg/L)	TS (mg/L)	TDS (mg/L)
1,260	315	292	135	22,010	22,170

ISEP Brine with 1% Ferric Chloride Surrogate

Oak Ridge National Laboratory developed a ferric chloride regeneration technique for ion exchange resin. This technique is particularly efficient for regenerating perchlorate-selective resin like A-530E, which is produced commercially by Purolite (Bala Cynwyd, Pa.). One of the main benefits of combining perchlorate-selective resin technology with ferric chloride regeneration is the potential to greatly reduce effluent volume (>99%) using conventional lead-lag ion exchange equipment. Demonstrating the ability to biodegrade mixed sodium chloride-ferric chloride brine will facilitate commercialization of the efficient ferric chloride regeneration technique by enabling disposal of effluent in facilities developed to address the sodium chloride brine effluent problem.

A representative surrogate ferric chloride solution was prepared based on three factors: 1) previous studies evaluating optimal ferric chloride regenerating solution composition; 2) the relative anion affinities for a perchlorate-selective resin, A-530E; and 3) a nominal groundwater composition consisting of 100 μg/L perchlorate, 40 mg/L of sulfate and chloride, and 15 mg/L of nitrate. The surrogate composition is shown in Table 6.13.

The test matrix for the mixed ferric chloride-sodium chloride brine is provided in Table 6.14. The feed for all conditions was prepared by addition of 1% by volume ferric chloride surrogate to La Puente sodium chloride brine (composite drum B) at the concentration specified in Table 6.14. The 1% concentration is representative of the relative effluent volume of ferric chloride vs. sodium chloride effluent generated from similar-sized drinking water treatment systems. The addition of ferric chloride surrogate to the sodium chloride brine resulted in an acidic feed solution that was fed to the R-1 without neutralization. Reactor pH was maintained by the addition of potassium hydroxide via the existing pH control system. The addition of the ferric chloride to the feed also increased the perchlorate concentration of the brine feed by a factor of 10. However, the nutrient ratio was still based on nitrate concentration.

Table 6.13
Ferric chloride surrogate composition

	Surrogate concentration (mg/L)		
Component	100%	5%	1%
HCl	182,250	9,113	1,823
$FeCl_3$	145,176	7,259	1,452
ClO_4^-	5,000	250	50
SO_4^-	4,000	200	40
Na^+	3,444	172	34
NO_3^-	500	25	5
HCO_3^-	500	25	5
TDS	340,871	17,044	3,409

Table 6.14
Test matrix for biodegradation of 1% ferric chloride surrogate in NaCl brine

#	Operating time (days)		HRT	Nutrient		Feed concentration (NaCl Brine)		
	Duration	Cumulative	(hr)	Ratio	(g/L)	Brine (%)	Cond. (mS/cm)	TDS (%)
1	5	5	12	3.5	2.8	50	49.4	3.43
2	1	6	12	3.0	1.9	40	42.1	2.88
3	2	8	12	4.35	2.8	40	42.1	2.88
4	11	19	16	3.0	1.9	40	42.1	2.88

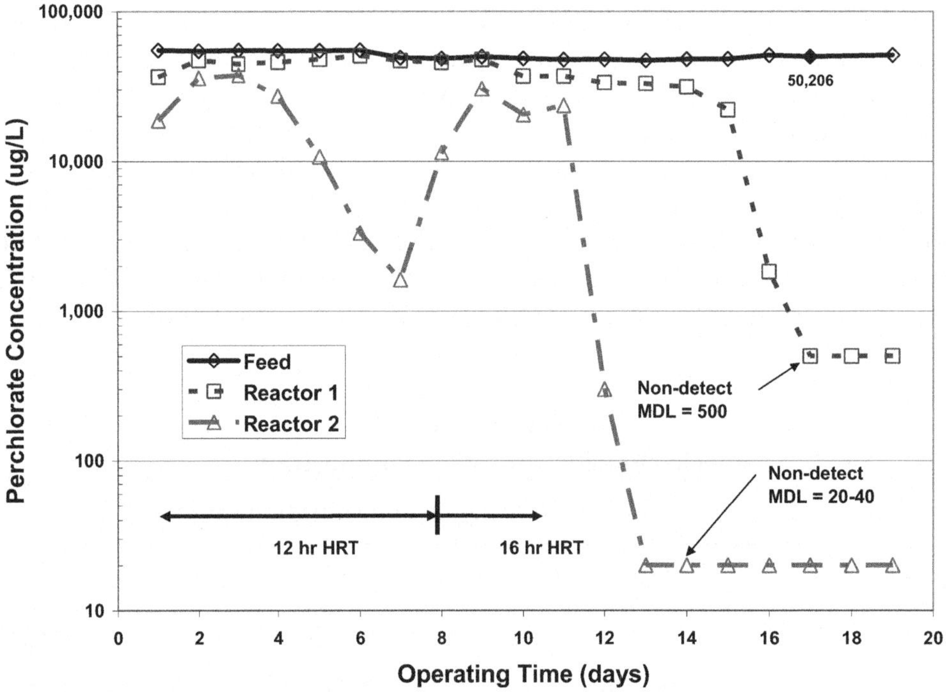

Figure 6.10 Perchlorate biodegradation performance – NaCl/$FeCl_3$ mixed brine

Figure 6.10 shows the perchlorate biodegradation results for the mixed brine feed solution. Only a slight reduction in perchlorate level was observed initially, but reducing the feed TDS to less than 3% by diluting the ISEP brine to 40%, and increasing nutrient concentration, resulted in over 97% perchlorate reduction on Day 7. After nutrient concentration was reduced and residence time was increased (Condition 4), a loss in reduction performance was observed. However, several mechanical problems contributed to the poor performance observed on Days 8 through 11. The problems included loss of flow from R-1 to R-2 and loss of temperature control, causing reactor temperatures to drop to 21.0 and 21.8°C respectively (from the 30°C setpoint). Even though very efficient perchlorate biodegradation had been demonstrated at 15°C, a sudden drop in temperature always results in a loss of activity and required several days for a culture to adapt. After resumption of temperature control at 30°C on Day 11, performance immediately improved and complete perchlorate reduction to below the method detection limit was obtained for the last 7 days of this test. This test demonstrated that, even though perchlorate feed concentration was significantly increased by the addition of ferric chloride brine and feed TDS was near 3%, complete perchlorate reduction could be achieved.

Pomona, California Brine Feed

The City of Pomona, California operates an ion exchange system to remove nitrate from their drinking water supply. This results in generation of spent brine with a nitrate concentration of approximately 7,000 mg/L. Three 5-gal containers of spent brine were obtained from the City. Table 6.15 provides a summary of analyses of the three individual drums. It is apparent by the anion concentrations that drum 2 was collected from a different part of the regeneration cycle than drums 1 and 3. The three drums were combined into one composite solution and diluted to 50%. This feed brine is characterized by having a much higher nitrate concentration and significantly lower perchlorate concentration than the brine from La Puente, California.

Table 6.16 provides the test matrix for the Pomona brine composite. The primary object of this test was to demonstrate complete, efficient nitrate reduction without particular emphasis on perchlorate reduction. As discussed previously, all nitrate and intermediate metabolite (nitrite) must be completely reduced to permit perchlorate reduction. The feed to the R-1 was switched from the previous feed (1% ferric chloride in sodium chloride brine) without cleaning or re-inoculating the reactors, or interrupting feed flow, again to minimize down time and demonstrate the adaptability of this culture.

Table 6.15
Pomona ion exchange brine composition

Drum	ClO_4^- (mg/L)	NO_3^- (mg/L)	SO_4^{2-} (mg/L)	pH	Conductivity (mS/cm)	TDS (%)
1	0.888	8,178	5,969	8.79	99.0	8.08
2	0.365	4,196	12,463	9.13	56.2	5.17
3	0.812	8,761	7,697	8.87	92.6	7.45
Composite avg.	0.688	7,045	8,710			6.90
50% Composite	0.344	3,522	4,355			3.45

Table 6.16
Test matrix for biodegradation of NaCl brine generated by the City of Pomona, California

	Operating time (days)			Nutrient		Feed Concentration		
#	Duration	Cumulative	HRT (hr)	Ratio	(g/L)	Brine (%)	Conductivity (mS/cm)	TDS (%)
1	3	3	16	1.5	5.3	50	46.7	3.22
2	5	8	12	1.5	5.3	50	46.7	3.22
3	9	17	16	1.5	5.3	50	46.7	3.22

Initially, complete denitrification was observed as shown in Figure 6.11. Within 2 days of reducing residence time from 16 hours to 12 hours, essentially all denitrification activity was lost. However, two separate mechanical problems resulted in low nutrient flow immediately before the upset and no nutrient flow for one day during the upset condition. This was the likely cause of the nitrate excursion. Nitrite was only observed in the reactors on days 11 and 12. Nitrite concentration was relatively high considering that it can be toxic. R-1 nitrite concentrations were 660 and 830 mg/L on days 11 and 12, respectively, and R-2 nitrite concentrations on the same two days were 2,020 and 1,870 mg/L. These results are consistent with previous tests that have showed significant nitrite generation upon initial nitrate reduction. Evidently, different metabolic pathways are employed for nitrite reduction to nitrogen than for nitrate reduction to nitrite. These pathways or enzymes must not be present at the onset of nitrate reduction but are quickly synthesized by the microbes when needed.

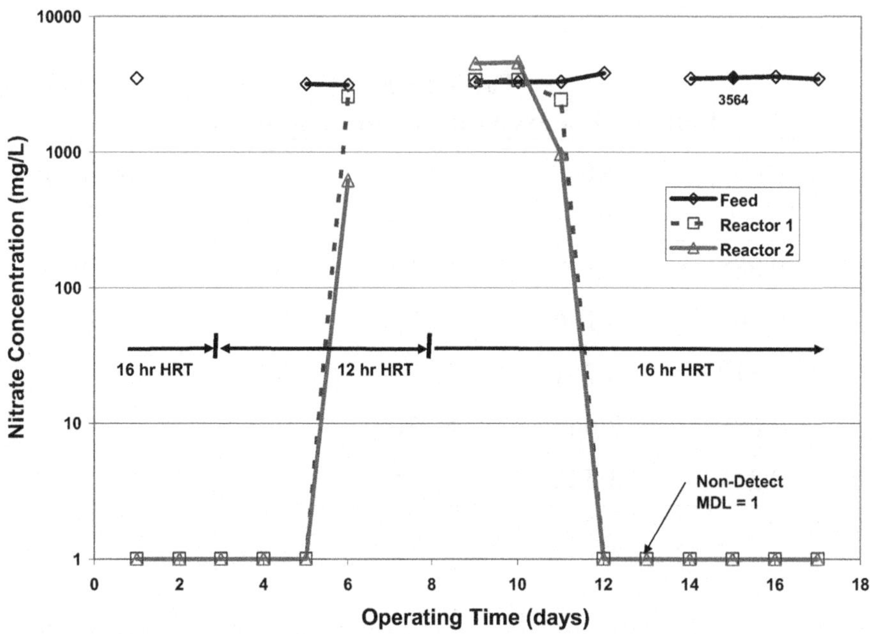

Figure 6.11 Nitrate biodegradation performance – Pomona brine

Figure 6.12 shows perchlorate reduction performance for the same time period. Initially, perchlorate was also being reduced, but not completely after day 5. Complete perchlorate reduction was obtained on day 13 and for the remainder of the test period. These results are consistent with previous tests performed by ARA that show that complete perchlorate reduction cannot be obtained until nitrite intermediate from nitrate reduction is completely reduced. This test demonstrated that complete perchlorate reduction can be achieved even when nitrate and sulfate concentrations are 10,000 times greater than perchlorate concentration.

In addition to demonstrating the reduction of perchlorate in mixed ferric chloride and sodium chloride brine as previously presented, the biodegradability of ferric chloride effluent alone is also important to the commercialization to this technology. A ferric chloride surrogate concentration of 5% was selected as a compromise to minimize treatment volume and maintain effluent TDS to less than 2 to 3% after neutralization. Table 6.17 provides the test matrix for biodegradation of ferric chloride surrogate that has been diluted to 5% of original strength. The nutrient ratio for this waste stream is based on perchlorate instead of nitrate because of the much higher perchlorate concentration. Startup problems resulted in very low reactor pH (less than 2.0) and damaged the microbial culture in the reactors. Therefore, it was necessary to re-inoculate the reactors for this test. Since the ferric chloride surrogate has a very low pH, it is important to determine the best approach to neutralize the ferric chloride solution without precipitating ferric hydroxide, which has low solubility near neutral pH and is difficult to redissolve into solution if precipitation has occurred. The approach used was to partially neutralize the 5% ferric chloride surrogate by addition of sodium hydroxide to increase the pH to approximately 2.4. Additional neutralization was accomplished in the R-1, which was maintained near pH 7.0 by addition of potassium hydroxide. This approach minimized ferric hydroxide formation and precipitation.

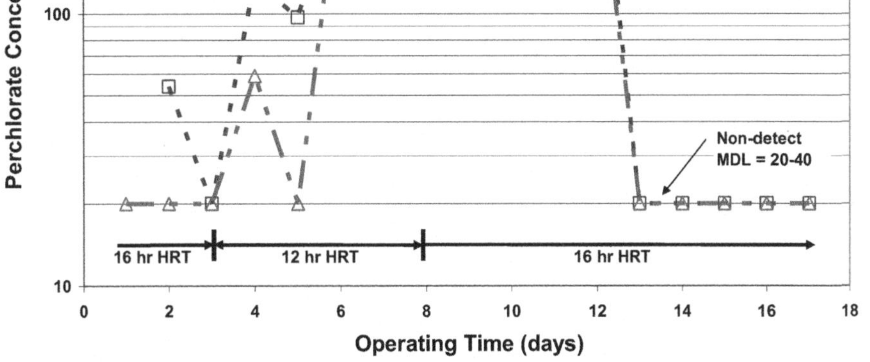

Figure 6.12 Perchlorate biodegradation performance – Pomona NaCl brine

Table 6.17
Test matrix for biodegradation of 5% ferric chloride surrogate

	Operating time (days)			Nutrient		Feed concentration		
#	Duration	Cumulative	HRT (hr)	Ratio	(g/L)	$FeCl_3$ (%)	Conductivity (mS/cm)	TDS (%)
1	3	3	16	5	1.3	5	22-24	1.41-1.55
2	6	9	16	8	2	5	22-24	1.41-1.55
3	1	10	16	6	1.5	5	22-24	1.41-1.55

The perchlorate biodegradation performance for the 5% ferric chloride test is shown in Figure 6.13. The feed perchlorate concentration is high compared to previous tests in this study. After flow was initiated, perchlorate reduction could not be maintained. Nutrient concentration was increased on day 4 (Condition 2) to improve performance. It appeared that the additional nutrient was beginning to improve performance by day 6. Also, on day 6 a nutrient modification was initiated. It appeared that the excess ferric chloride in the reactors was causing some additional flocculation. It is also known that an effective technology for nutrient removal (i.e., phosphate reduction) is precipitation by ferric chloride. To negate the impact of phosphate removal by ferric chloride, additional phosphate and micronutrient were added to the nutrient mix beginning on day 6. The combination of higher nutrient ratio and additional phosphate were very effective in promoting complete perchlorate reduction on day 7 and for the remainder of this test. While this short test did not permit optimization, it is likely that with proper phosphate and micronutrient supplement for this feed matrix, nutrient ratio of 3 to 4 could be obtained.

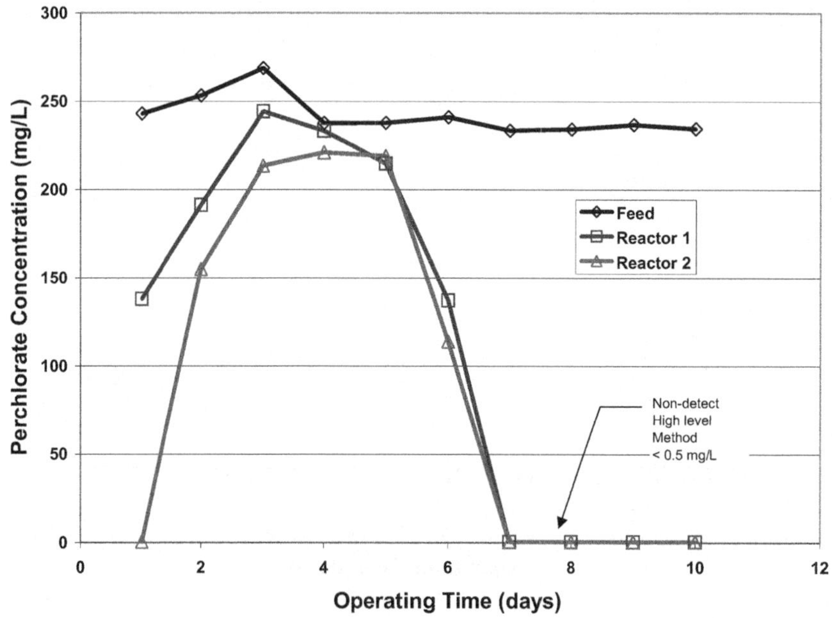

Figure 6.13 Perchlorate biodegradation results – ferric chloride brine

Biodegradation Bench Testing of Undiluted Brine

Brine used as the feed source during these tests was collected from the ISEP process at La Puente. The collection location of the brine was after the addition of rinse water and acid, but before effluents from other demonstration processes were returned to the brine discharge line. Therefore, the brine is representative of the total ISEP process discharge stream. Preliminary batch and sequencing batch reactor studies were conducted that were not part of this project; however, a brief summary of these results is included because results contributed to determining flow reactor operating parameters. In some cases, due to limited brine availability at desired salt concentrations, brine was amended with water softening salt (99.8% NaCl). Table 6.18 summarizes the media used for the undiluted brine studies.

Biological Sequencing Batch Reactor Bench Equipment Description and Operation

A 6-liter batch reactor system was used to evaluate the biodegradability of undiluted brine (Figure 6.14). Initially, undiluted brine and CSB nutrient were added directly to the reactor and the pH of this mixture was adjusted to approximately 7.7. The reactor was inoculated with 200 mL of 6.9% TDS inoculum, and 200 mL of 10,000 mg/L NO_3^- inoculum that are maintained by ARA at their Panama City research facility. This reactor system was sealed from air infiltration and mechanically agitated. The pH was controlled at 7.0 by addition of 15% potassium hydroxide. Reactor temperature was maintained at 35°C. The ORP was monitored daily.

The reactor was operated in a manner similar to a sequencing batch reactor (SBR) system. When perchlorate was reduced to below the detection limit (~50 µg/L in this media), agitation, pH control, and temperature control were halted and suspended solids (biomass) were allowed to settle for 2 hours. During the settling period, a new batch was prepared by adding 40 mL of nutrient to 4 L of La Puente brine and by adjusting the pH to 7.6. After the settling period, four liters of supernatant was decanted from the reactor and the fresh brine and nutrient mixture was added. Agitation, pH control and temperature control was then resumed. The 2-L heal left in the reactor provided inoculation for the next cycle of operation.

Nutrient formulation was varied to improve perchlorate reduction in undiluted brine. Variations consisted of increasing phosphate, increasing micronutrients, or increasing both compared to typical formulations.

Table 6.18
Undiluted brine properties

		Total dissolved solids (%)		Composition (mg/L)		
Study	Source	As received	Amended	ClO_4^-	NO_3^-	SO_4^{2-}
SBR	Drum	6.2	No	4.8	2,316	2,762
Flow	Tank truck	4.7	6.2	2.7	1,655	2,140
Flow	Drum	5.1	No	4.2	3,630	556

Figure 6.14 Sequencing batch reactor

Results of Undiluted Brine Sequencing Batch Reactor Tests

SBR studies on undiluted brine showed that nitrate was typically biodegraded within 24 hours and nitrite within 48 hours. Complete perchlorate reduction was obtained in 7 to 10 days. It was anticipated that performance would gradually improve through each cycle due to a gradual increase in biomass; however, a significant improvement in performance was not observed. It was suspected that biomass was not increasing as anticipated. VSS analysis of the supernatant before and after the 2-hour settling period confirmed that the biomass concentration decreased by only 8% after settling. Therefore, the biomass was not being efficiently retained in the reactor from one cycle to the next cycle.

There were two positive results from initial SBR studies: 1) perchlorate could be reduced to below the EPA 314.0 MDL (0.53 μg/L); and 2) initial perchlorate reduction rates were typically very high. The initial reduction rates during the first 2 days of operation were typically 500 to 1,000 mg/L-day. If conditions could be optimized to maintain this rate, complete perchlorate reduction could be achieved in 3 to 4 days versus 7 to 10 days. Short cycle time for SBR operation would give confidence that undiluted brine could be biodegraded in a continuous flow reactor system.

It was theorized that by optimizing conditions for nitrate reduction followed by perchlorate reduction, higher reduction rates could be maintained throughout the membrane bioreactor (MBR) cycle. Adding excess micronutrients and controlling the pH of the system at specific set points achieved the best perchlorate reduction rates. Figure 6.15 is an example of perchlorate reduction performance that was obtained using the sequencing batch reactor system.

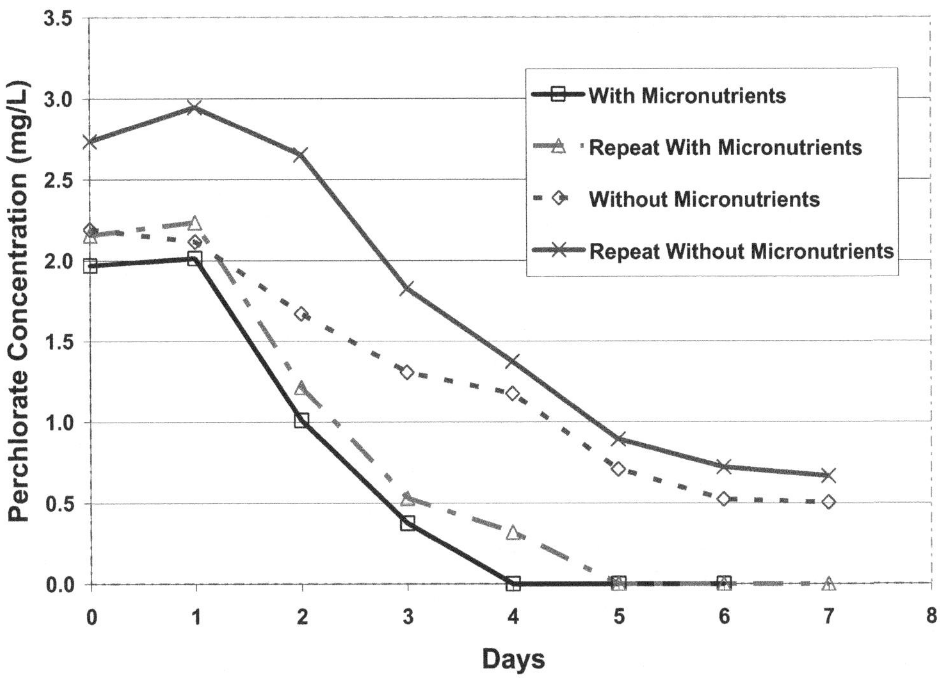

Figure 6.15 SBR biodegradation results

During the studies conducted with micronutrients, the micronutrient concentration was increased to four times the standard formulation in an attempt to maintain availability throughout the perchlorate reduction period. The pH was maintained above 7.1 initially to create favorable denitrification conditions. When the nitrate and nitrite were reduced to below detection limits (on day 2), the reactor was amended with additional micronutrients. Also, the pH was reduced to 7.1 and controlled at less than 7.1 for the duration of the test to maintain favorable conditions for perchlorate reduction. This test resulted in the most rapid perchlorate reduction of any of the SBR tests. Complete perchlorate reduction was achieved in only 4 days. A repeat test resulted in similar performance, with complete perchlorate reduction taking in 5 days.

The conditions for the studies conducted without additional micronutrients were identical to the two preceding tests except that the reactor was not amended with micronutrients after nitrate reduction was complete. Performance was not as good as the previous two tests indicating that the extra micronutrients may have played an important role in maintaining high perchlorate reduction rates. Perchlorate was not completely reduced during the 7-day test period.

Biological Continuous Flow Reactor Bench Equipment and Operation Description

The batch reactor studies provided evidence that perchlorate biodegradation could be optimized by addressing nitrate and perchlorate as separate reactions occurring in series. This may be accomplished in a two-stage flow reactor system by operating each reactor independently and tailoring conditions for nitrate and perchlorate reduction in the first- and R-2s respectively. In addition, since perchlorate reduction is much more difficult to accomplish than nitrate reduction, the R-2 for this test was larger than the R-1 by a factor of 2.4, (i.e., 6-L R-2 and 2.5-L

R-1). This enabled a much longer residence time in R-2. Based on the SBR studies, micronutrient concentration was increased. Figure 6.16 is a picture of the two-stage, bench reactor system. The pH in R-1 was maintained at 7.0 or greater by addition of 15% potassium hydroxide. The pH in R-2 was maintained at 7.15 or less by addition of 1N hydrochloric acid fortified with additional micronutrients. A limited quantity of 6.2% TDS undiluted brine from La Puente was available. Therefore, brine used for this system consisted of 4.7% TDS undiluted brine from La Puente amended with water softening salt to increase the TDS to 6.2%.

Biological Continuous Flow Reactor Inoculation and Test Matrix

The reactor system was configured and the R-1 was inoculated using water from the batch reactor, which became the R-2. After inoculation, flow was initiated at a residence time of 24 hours in the R-1. This flow rate resulted in a residence time for the R-2 of approximately 58 hours. Table 6.19 provides a summary of the test conditions.

Biological Continuous Flow Reactor Results

Initial performance was poor as evidenced by almost no perchlorate or nitrate reduction. On January 15, 2004, a microscopic inspection of both reactors revealed a very sparse bacteria population. Evidently, bacterial growth rate had been slowed to the point that a healthy culture could not be effectively maintained at a 24-hour residence time. Flow to the reactor system was suspended and 1-L of water was removed from R-1 and replaced with 1-L of inoculum. Both reactors were spiked with nutrient. To retain biomass in the reactor system, a gravity clarifier was used to clarify effluent from R-2 and return biosolids to R-1. Flow was resumed on January 16, 2004. Figure 6.17 shows a dramatic improvement in performance as a result of these corrective actions.

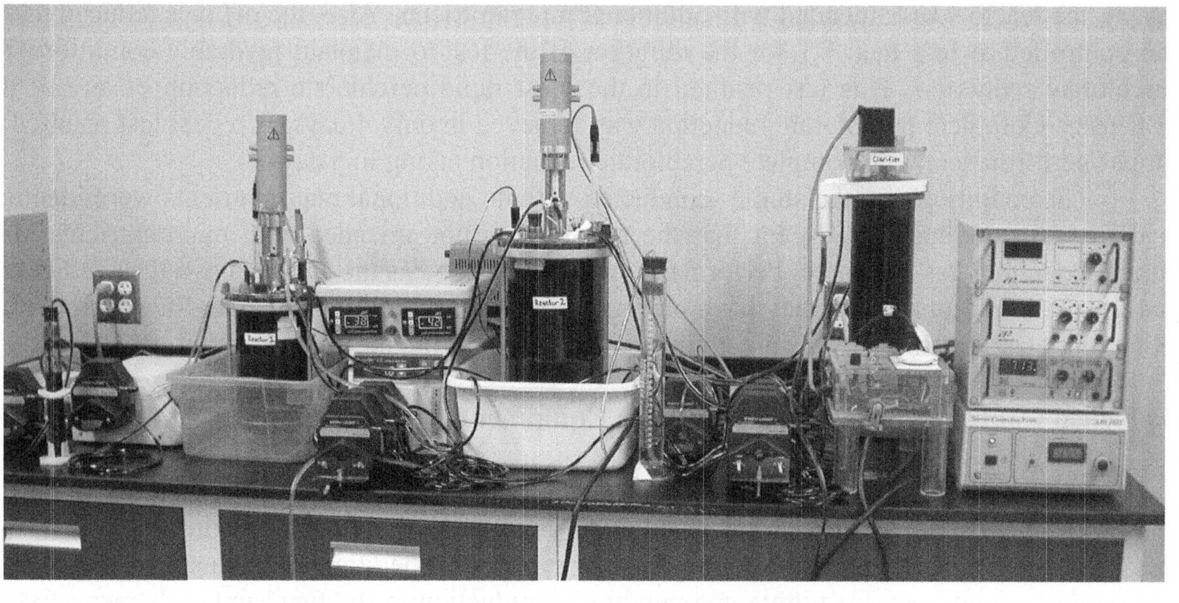

Figure 6.16 Two-stage bench reactor system – undiluted brine

Table 6.19
Test matrix – undiluted brine flow reactor

Date	HRT (hr)*	Nutrient (vol %)	Recycle rate (%)†	Brine feed
1/6/04	24	1.00	0	Amended 6.2 TDS
1/12/04	24	0.51	0	Amended 6.2 TDS
1/15/04	Batch	Spiked	0	Amended 6.2 TDS
1/16/04	24	0.51	100	Amended 6.2 TDS
1/23/04	24	0.51	50	Amended 6.2 TDS
1/26/04	24	0.77	50	Amended 6.2 TDS
1/29/04	24	0.51	50	Amended 6.2 TDS
2/4/04	24	0.51	50	5.1 TDS undiluted
2/6/04	24	1.25	50	5.1 TDS undiluted
2/13/04	24	1.25	0	5.1 TDS undiluted
2/20/04	Shut down reactor system.			

* Based on brine feed rate to R-1
† As % of brine feed

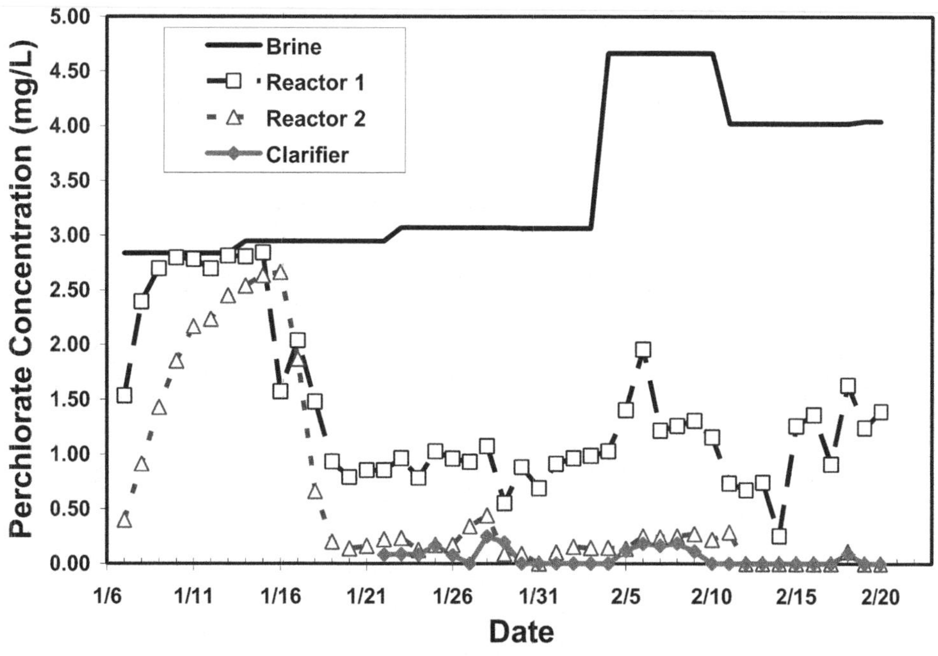

Figure 6.17 Flow reactor performance – undiluted brine

Complete perchlorate reduction was obtained in the clarified effluent until the feed was changed to undiluted, unamended 5.1% TDS brine from La Puente on February 4, 2004. This brine contained significantly more nitrate (approximately 3,600 mg/L vs. 1,650 mg/L) than the previous brine. After the nutrient feed concentration was adjusted on February 6, 2004 for the increased nitrate concentration in the new brine feed (see Table 6.19), complete perchlorate reduction was again observed in a flow reactor system treating undiluted brine.

To confirm the benefit of the clarifier, recycling was discontinued on February 13, 2004. Perchlorate concentration in R-1 gradually increased. Also, after one week, biomass concentration as determined by VSS, decreased by 52% in R-1 (from 920 to 443 mg/L) and by 15% (from 702 to 615 mg/L) in R-2. This trend would likely continue and impact perchlorate reduction performance.

PILOT-SCALE FIELD DEMONSTRATION MATERIALS AND METHODS

Transportable Biodegradation System Description

The ARA transportable biodegradation system is modeled after the simple, bench-scale, two-stage reactor biodegradation process patented by ARA (Patent Number: 6,077,432). Reactors are operated as continuous stirred tank reactors (CSTRs) that support the suspended-growth biodegradation of nitrate and perchlorate.

Additional capabilities of the transportable system include unattended operation, fail-safe design for power interruptions, overflow protection, simple, local control systems (pH, temperature, and flow), and the ability to operate in both batch and continuous flow modes of operation. The system is designed to treat a wide range of waste streams and to function over a wide range of operating conditions. The specific goal of the La Puente field demonstration was to duplicate the successful perchlorate reduction in ISEP brine that was demonstrated in the lab study. Table 6.20 lists the design basis of the transportable system as constructed.

Table 6.20
Transportable system design basis

	Minimum	Maximum
Flow rates		
Brine (L/min)	0.3	3.0
Dilution water (L/min)	0.3	3.0
Nutrient (mL/min)	1	50
Caustic (ml/min)	50	-
HRT (hr)*	3	24
Temperature (°C)	15	40
Brine concentration (%)	10	100

* Reactor Working Volume = 275 gals each (~1040 liters)

The biodegradation system is housed in an 8.5-foot wide by 20-foot long by 8-foot high, enclosed Haulmark® cargo trailer. The trailer has an average payload capacity of 8,700 pounds and comes equipped with tandem 6,000-pound axles, 4-wheel electric brakes, and a 2-5/16-inch hitch. Trailer components also include rear cargo doors, 4-foot side door, 3/4-inch plywood walls, insulated walls and ceilings, 1-1/2-inch layer plywood floors, 2 roof vents, AC/heating system, 4 fluorescent lights, 240-volt, 50-amp breaker box, heavy-duty leveling jacks, front and rear, and sprayed-in, non-skid flooring. Figure 6.18 shows the trailer being set up at La Puente.

The transportable biodegradation system consists of two 300-gal tanks (R-1 and R-2); brine, dilution water, and nutrient feed systems, and local controls for pH and temperature. Brine and dilution water are simultaneously metered directly into the R-1. Brine from the ISEP process is delivered to a 4,000-gal Baker tank that provides an equalized, consistent feed source (Figure 6.19). Brine is delivered to R-1 via a Milton Roy "MacRoy" (Ivyland, Pennsylvania) diaphragm metering pump. Potable water is used for brine dilution. Dilution water flow rate to R-1 is monitored and controlled by a rotameter. A dilution water pressure is regulated to maintain constant system pressure and enables accurate flow control. Figure 6.20 shows both the brine and dilution water feed systems. As an additional verification that proper dilution of the brine is being maintained, the conductivity of R-1 is monitored.

Figure 6.18 Pilot trailer on site w/4,000-gal Baker tank

Figure 6.19 Brine feed equalization tank

Figure 6.20 Brine feed pump and water feed system

The pH is monitored and controlled in each reactor by Yokogawa model PH150 'EXAxt' pH and ORP controller (Figure 6.21). Caustic (15% potassium hydroxide) is delivered to each reactor via Milton Roy LMI metering pumps (Ivyland, Pennsylvania) (Figure 6.22). Nutrient is metered to R-1 only by a peristaltic pump. Small March magnetic drive pumps provide agitation by recycling reactor contents through an eductor located near the bottom of each reactor. R-1 flows by gravity to R-2, which, in turn, is discharged by gravity to a sump located adjacent to the trailer. The sump pump periodically transfers biodegraded effluent into the main brine discharge sump. Figure 6.23 is a schematic of the reactor configuration.

Field Demonstration Setup and Inoculation

ARA personnel prepared the R-1 for inoculation on October 1, 2003. Brine and dilution water were charged into the R-1 to achieve a conductivity of 32 mS/cm (approximately 2.1% TDS) and raise the reactor level above the reactor heating tape elements. Reactor contents were circulated overnight and the temperature increased to 85°F (29°C) and pH 7.2. On October 2, 10 mL/L of CSB nutrient was added to the reactor. After the pH returned to 7.2, the reactor was inoculated using 1 gal of inoculum that was adapted to high TDS media. Initial reactor samples were collected. Performance was monitored over the next several days and flow to the system was initiated on October 6, 2003.

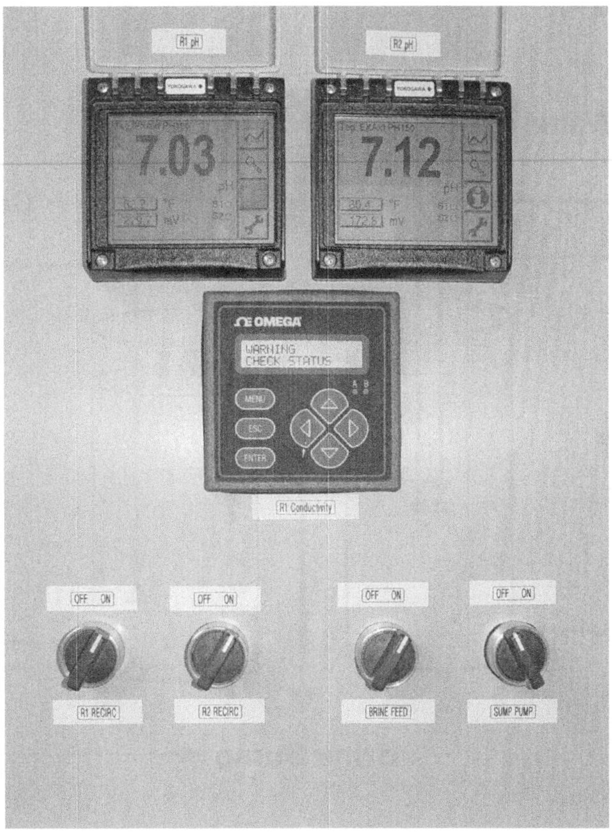

Figure 6.21 pH controllers and control panel

Figure 6.22 Caustic metering pumps

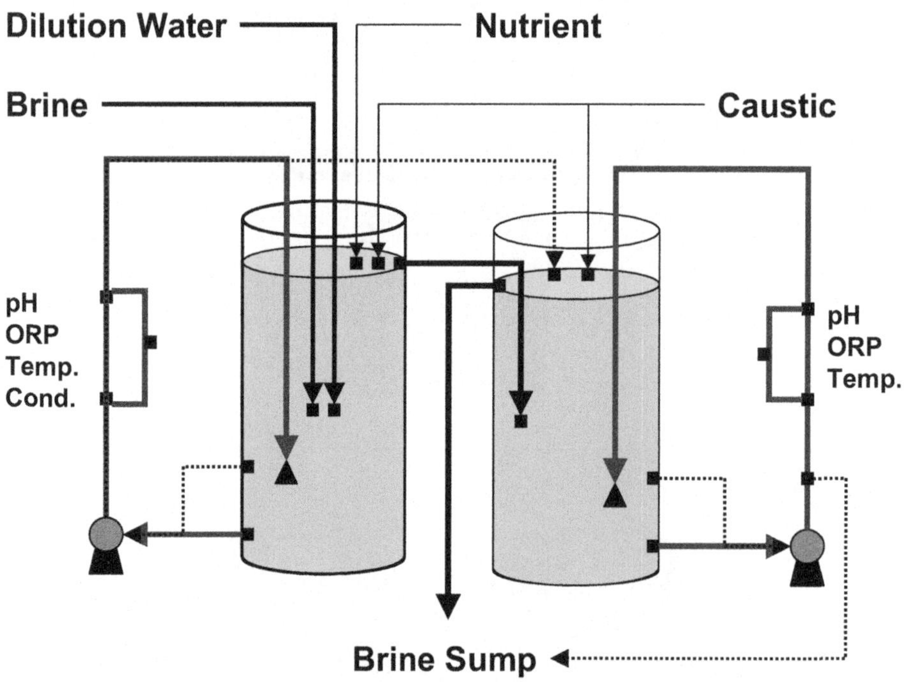

Figure 6.23 Schematic of reactor configuration

Data Collection, Sampling, and Analysis

Data collection/log sheets were developed for recording the reactor status, process set points, and field analytical data. In addition, an O&M checklist and "field notebook" were provided for operators from ARA or Carollo.

For the transportable biodegradation system, there are three primary samples points: the undiluted brine feed stream, R-1, and R-2. Each sample is clearly labeled and the label firmly affixed with the following information: sampling point; R-1, R-2, or Brine Feed; date (mm/dd/yy); and time (24-hour clock). Each reactor was sampled through sampling valves located in the sensor/recirculation loop. After collecting samples in 10 mL test tubes, the tubes were tightly closed, labeled, and immediately placed in a test tube rack in the freezer on site. A similar procedure was followed for collecting brine samples from the brine feed pump suction line. Preserved samples were shipped frozen via overnight delivery to ARA's Panama City laboratory for analysis of anions and perchlorate using analytical methods described previously. The only field analysis performed was for nitrate and nitrite with commercially available test strips (Quick Dip 5-N-1 test strips, Jungle Laboratories, Cibolo, Texas). Samples were collected two or three times per week and shipped once per week. This caused up to a two-week lag in obtaining analytical results. The inability to obtain near real-time performance feedback severely limited the capability to optimize system operating parameters. Two sets of samples were collected and analyzed by Weck Laboratory, a California NELAC certified lab.

Brine Feed Analysis

Obtaining and maintaining a consistent, representative quality of brine feed was a priority of this pilot demonstration. The sample point was the same location as for bench test samples (Figure 6.2). This location was immediately upstream of a 4-inch ball valve used to provide approximately 4 to 6 psig of backpressure to the ISEP system. This was not enough backpressure to fill the 4,000-gal Baker storage tank used for brine feed equalization. Additional backpressure (8 to 10 psig) was applied during attended operation to partially refill the Baker tank and maintain sufficient margin and head for safe operation. This caused the brine quality to change with time. Figure 6.24 shows the TDS and perchlorate analyses with time. The TDS and perchlorate concentrations appeared to be proportional to one another. Initially, TDS was higher than anticipated (~8%), but decrease to an average near 6% during the middle three weeks of operation. TDS concentration became erratic and lower (4 to 5%) during the last 2 weeks of operation.

Figure 6.25 shows the nitrate and sulfate concentrations in the undiluted brine compared to the perchlorate concentration. With the exception of a couple of excursions, sulfate concentration paralleled perchlorate concentration and gradually decreased with time. Except for high initial values, nitrate concentration remained relatively constant at approximately 2,000 mg/L.

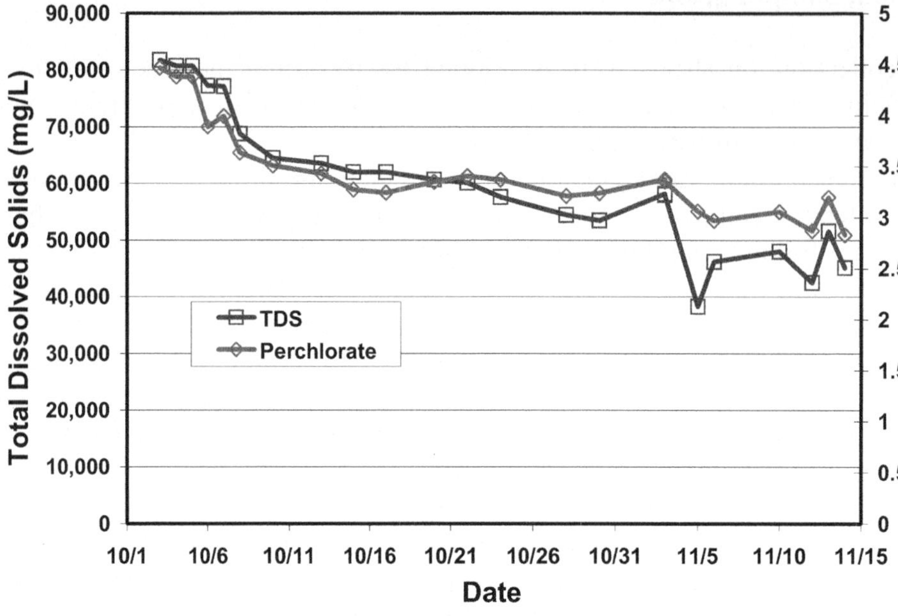

Figure 6.24 Undiluted brine TDS and perchlorate concentration

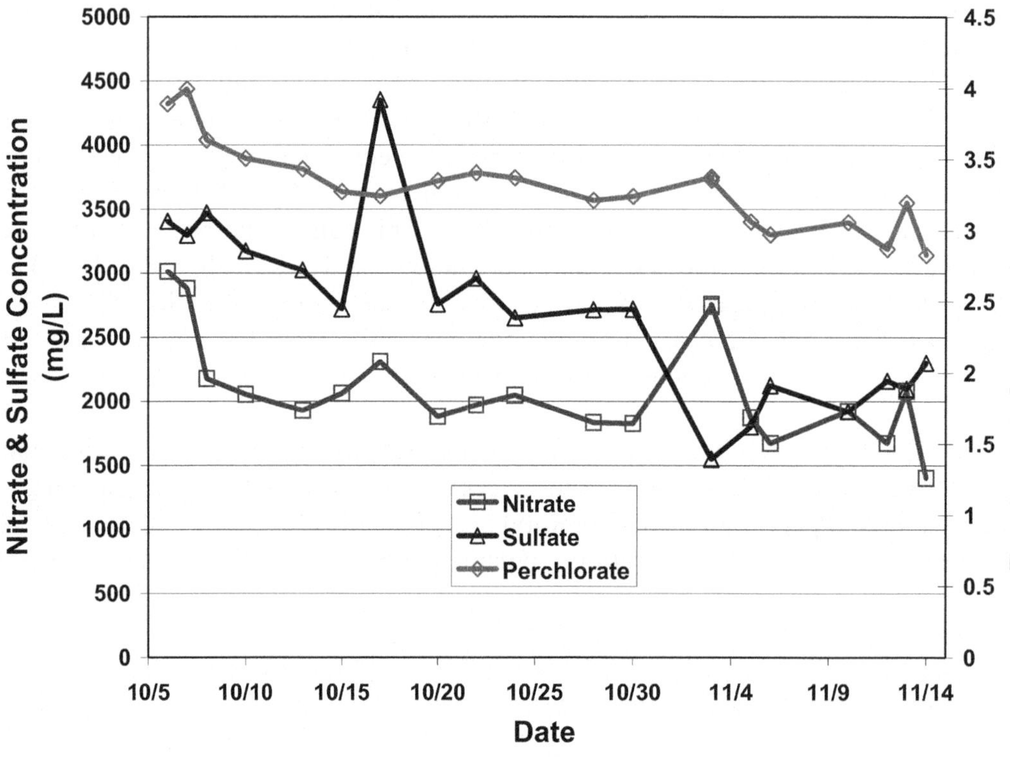

Figure 6.25 Undiluted brine nitrate and sulfate concentrations

Brine Concentration in Reactor Feed

Brine was not pre-diluted with water before being pumped to the reactors. Since brine and dilution water were metered directly into the R-1, several variables created errors in maintaining the target brine concentration. The high bicarbonate concentration of the brine led to the formation of carbon dioxide in the brine. A combination of factors including the addition of acid to the brine in the ISEP process, relatively cool brine storage temperature with up to 10 feet of head pressure, and rapid depressurization in the suction line of the feed pump, caused the formation of CO_2 gas bubbles in the brine feed. Gas in the suction line affected both calibration curves and metering accuracy. Dilution water was metered by a simple rotameter and needle valve. A pressure regulator maintained essentially constant pressure at the rotameter, but this did not eliminate fluctuations in dilution water flow rate over time.

The conductivity of R-1 was monitored to check brine dilution. However, this method also had limited accuracy and reliability. The conductivity probe itself was fouled by reactor water and changing ion concentrations also impacted conductivity measurements. If the concentration of brine delivered to R-1 is calculated based undiluted brine TDS and R-1 conductivity, inconsistent results were obtained (i.e., actual reactor concentrations are sometimes higher than calculated feed concentrations). Sulfate was relatively non-reactive in the reactors and, therefore, was used as a "tracer" compound to normalize results and calculate the actual brine concentration delivered to the reactor. Figure 6.26 shows the results of the two methods used to confirm actual brine concentration. R-1 sulfate concentration was used to normalize brine feed concentration for the purpose of evaluating perchlorate reduction performance.

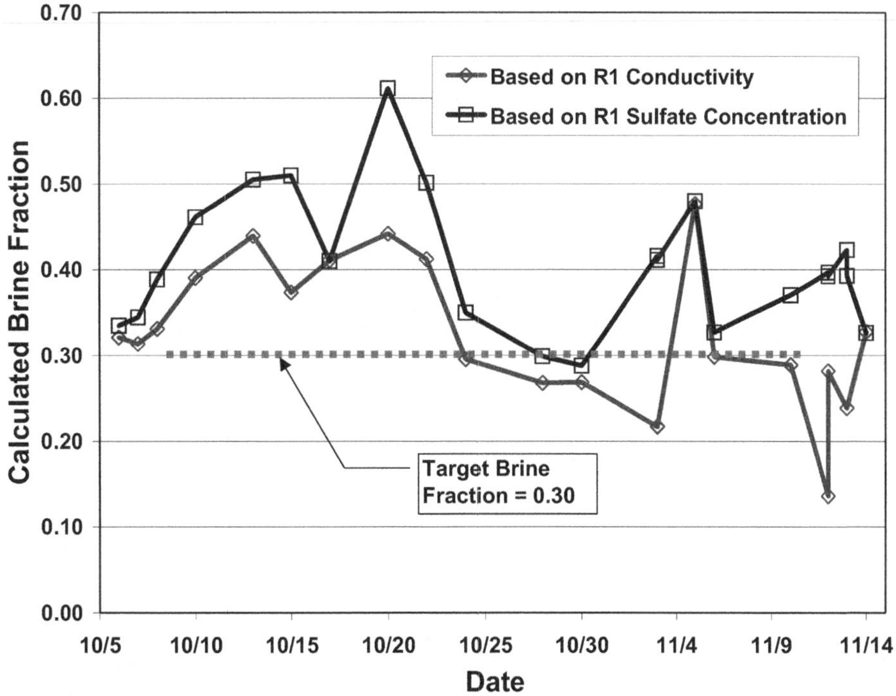

Figure 6.26 Calculated brine feed concentration

PILOT-SCALE FIELD DEMONSTRATION RESULTS AND DISCUSSION

Table 6.21 shows the test matrix and reactor operating parameters. Nutrient concentration was based on the assumption that the nitrate in undiluted brine was 3,000 mg/L and the brine was diluted in the reactor to 30%. In reality, nitrate concentration was 3,000 mg/L initially, but stayed around 2,000 mg/L for most of the duration of this test (Figure 6.25). This was partially offset by the actual brine concentration in the reactor, which was usually above 30% and fluctuated to concentrations as high as 60% (Figure 6.26). The nutrient ratios in Table 6.21 were calculated based on the actual nutrient flow rate and the assumption that the brine concentration would stay constant at 30%.

Near complete nitrate reduction was observed during the field demonstration test. Figure 6.27 shows that two nitrate excursions occurred. The excursion on October 17, 2003 may have been caused by a combination of high brine feed fraction (near 50%, Figure 6.26) and the relatively short residence time (12 hours). Residence time was increased on the October 17. A second excursion occurred November 17, 2003. Previous studies on similar high-TDS media showed that perchlorate reduction performance could be improved by operation at higher pH to maximize nitrate reduction rates. However, that was not the case with this media and the increase in pH indirectly caused a small nitrate excursion.

Figure 6.28 shows the perchlorate reduction performance. While some perchlorate reduction was evident throughout the test period, near-complete reduction was not observed until November 5, 2003. The inability to optimize pH and nutrient during the course of testing, along with the fluctuating properties of the brine feed and higher than desired TDS contributed greatly to less than anticipated performance during the first four weeks of this demonstration. The 10 days leading up to the period of good performance (October 24 to November 3) was characterized by relatively stable feed brine concentration near the target value of 30% (Figure 6.26). The lowest perchlorate value obtained was 80 µg/L, which is near the method detection limit in this matrix. Increasing pH on November 17, 2003 also had an adverse effect.

Table 6.21
Test matrix for the pilot biodegradation of NaCl Brine from the ISEP process

	Operating time (days)			HRT	Nutrient			
#	Date	Duration	Cumulative	(hr)	Ratio	(Vol. %)	Brine (%)	pH
1	10/6/03	1	1	16	3.1	0.42	30	7.0
2	10/7/03	8	9	12	3.1	0.42	30	7.0
3	10/15/03	2	11	12	3.1	0.42	30	7.0
4	10/17/03	28	39	16	3.6	0.49	30	7.0
5	11/14/03	5	44	16	3.6	0.49	30	7.3
6	11/19/03	1	45	16	3.6	0.49	30	7.6
7	11/20/03	5	50	16	3.1	0.42	30	6.7

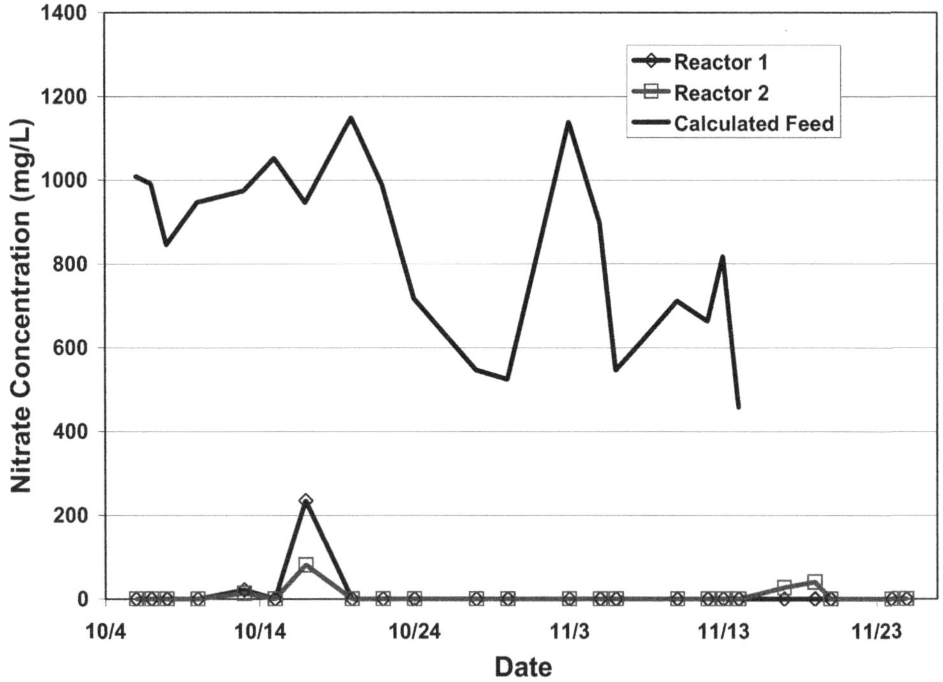

Figure 6.27 Nitrate reduction performance from pilot test

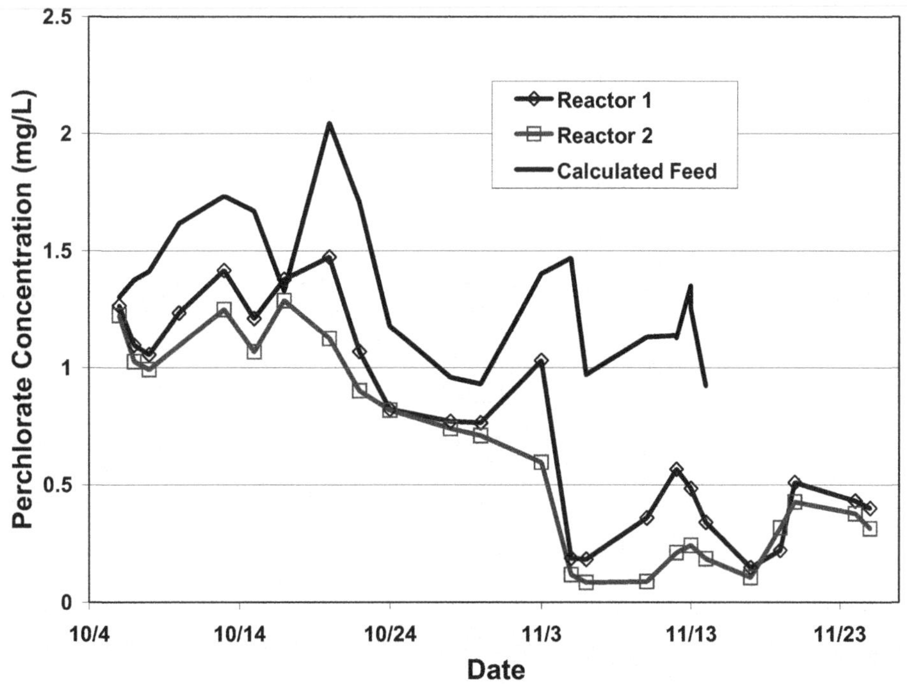

Figure 6.28 Perchlorate reduction performance from pilot test

NELAC Certified Lab Analysis Results

Two set of samples were collected for analysis by an external-certified laboratory during the field demonstration. Standard method protocols regarding sampling collection, chain-of-custody and preservation procedures were followed. The first set of samples was collected on November 14, 2003, and the second on November 25, 2003. Table 6.22 highlights the results from these analyses. Results were reported with no flags for data qualifiers and the quality assurance and quality control report for the these sample sets indicates acceptable results for blanks, spikes, and duplicates for all of the analyses performed as per the method.

Table 6.22
Weck laboratory analysis

	Results (11/14) mg/L	Results (11/25) mg/L	Dilution factor	Reporting limit	Method
Brine					
Total alkalinity ($CaCO_3$)	3,400		1	2.0	SM 2320B
Carbonate alkalinity ($CaCO_3$)	ND		1	2.0	SM 2320B
Bicarbonate alkalinity (HCO_3)	4,200		1	2.0	SM 2320B
Hydroxide alkalinity ($CaCO_3$)	ND		1	2.0	SM 2320B
Perchlorate (μg/L)	2,900		50	150	EPA 314.0
Reactor 1					
Perchlorate (μg/L)	ND		10	30	EPA 314.0
Reactor 2					
Total alkalinity ($CaCO_3$)	2,700		1	2.0	SM 2320B
Carbonate alkalinity ($CaCO_3$)	ND		1	2.0	SM 2320B
Bicarbonate alkalinity (HCO_3)	3,300		1	2.0	SM 2320B
Hydroxide alkalinity ($CaCO_3$)	ND		1	2.0	SM 2320B
BOD	1,400	1,500	1	10	SM 5210B
COD	3,000	3,300	4	20	EPA 410.4
Total phosphorus	18	10	100	5.0	EPA 365.3
Total solids	19,000	16,000	1	10	EPA 160.3
Total suspended solids	140	360	1	5	EPA 160.2
Total volatile solids	2,300	3,300	1	5.0	EPA 160.4
Total Kjeldahl nitrogen	130	140	1	0.1	SM 4500
Perchlorate (μg/L)	ND	410	10/20	30/60	EPA 314.0

Perchlorate analysis of the feed water and reactor samples is in general agreement with analyses conducted by ARA. On November 14, 2003, the feed brine contained 2,900 µg/L perchlorate (2,828 by ARA analysis). Weck laboratory reported perchlorate at non-detect in reactors 1 and 2 on that same day with a 30 µg/L detection limit. ARA results on split samples that were immediately frozen (vs. preserved on ice) were 340 and 185 µg/L for reactors 1 and 2, respectively. The very high bicarbonate alkalinity buffered the pH of the brine and could result in excessive chemical consumption if significant pH adjustment is necessary. The very high BOD and COD results provide further evidence that the reactors were fed excess nutrient.

SUMMARY AND RECOMMENDATIONS

High TDS conditions reduced biomass production and decreased the reduction rates for both nitrate and perchlorate. Therefore, it is necessary to maintain modest brine concentrations (2 to 3%) and moderately long residence times (12 to 16 hours) in suspended-growth, stirred-tank-reactor systems. Brine biodegradation conducted under this project did not address recovery and recycle of biomass to improve performance. However, other ongoing studies have shown that suspended growth processes, with mechanisms to retain biomass, significantly improved performance and enabled the biodegradation of brine at much higher concentrations. Biomass recovery would also reduce disposal premiums that may be levied on treated effluent.

Brine biodegradation system upsets can be caused by mechanical failure that impacts feed or operating conditions. The impact of system upsets is mitigated by several measures, which include: 1) appropriate alarm and alarm notification; 2) routine monitoring, sampling, and analysis; 3) on-line nitrate/perchlorate analysis; 4) appropriate surge capacity; and 5) appropriate permit provisions. The first four measures provide an opportunity to avoid an upset condition. In the event a serious upset occurs that requires re-inoculation or restart of the system, permit provisions must provide for the ability to discharge untreated brine for one to two weeks. During over 6 years of continuous operation of the ARA system at ATK-Thiokol, since December 1997, only one upset has occurred requiring re-inoculation.

The detection limit of the current analytical method for perchlorate, EPA-314, is limited due to the high TDS of the brine matrix and potential for organic interference. While regulatory guidelines for discharge of treated brine have not been firmly established, it is assumed that some level, at or above the current Public Health Goal (PHG) of 6 µg/L, will be established for California. New methods for perchlorate analysis are being developed that can accommodate high TDS and organics. The high TDS wastewater discharged from the ARA treatment facility in Kansas limits EPA-314 to an MDL of 20 µg/L and is also plagued by organic interference. Since the Kansas Department of Health and the Environment (KDHE) limits perchlorate discharge to less than 100 µg/L, this does not typically present a problem. However, organic interference has generated a number of false positive results. Retests using the Liquid Chromatography-Mass Spectrometry (LC-MS) method 8321A reported perchlorate below the method detection limit of 2 µg/L.

Discharge of treated brine is a concern due to high TDS. Although discharge to the brine line is acceptable, surcharges may apply for suspended solids, chemical oxygen demand (COD), and salt content. Discharge pH may have to be monitored and controlled to minimize the scaling potential of the discharge based on Langelier Saturation Index (LSI). Hydrochloric acid is currently being added to untreated brine discharged from La Puente to maintain a proper LSI.

The anaerobic nature of the water does not present a problem since sewer line wastewater is typically anaerobic.

Based on the results of this study, it is recommended that perchlorate biodegradation of this brine be evaluated using a biodegradation system that recovers and reuses biomass. This can be accomplished by using technologies such as sequencing batch, which was demonstrated in this study, fixed film bioreactors, or adding a clarifier or a membrane filtration step. Biomass concentration and reuse should improve perchlorate biodegradation performance in addition to minimizing effluent disposal and system operation costs.

CHAPTER 7
PERCHLORATE AND NITRATE THERMAL BRINE TREATMENT

INTRODUCTION AND OBJECTIVES

Although thermal treatment of perchlorate in NaCl brines can be accomplished (Li and Coppola 2000), the energy costs due to evaporation and brine concentration requirements can make this process cost prohibitive. With availability of perchlorate selective resins and an alternative regeneration option using $FeCl_3$/HCl as previously described (Gu et al. 2001), significantly reduced brine volume is generated. Thus, the thermal treatment of $FeCl_3$/HCl brine can be a cost effective strategy to deal with this waste stream. A unique aspect of this task is to use a relatively low-temperature process to completely destroy perchlorate in spent $FeCl_3$/HCl brine.

The objectives of the thermal treatment tests are to: 1) validate the decomposition temperature of perchlorate residuals in ferric chloride brine; 2) determine the decomposition rate at the validated decomposition temperature; and 3) examine the feasibility of HCl recovery and reuse. A ferric chloride brine surrogate (same formulation used for biological testing) was used for all decomposition tests.

MATERIALS AND METHODS

The surrogate was prepared and 0.5 mL of the solution was added to seven test tubes, one of which was diluted to 10 mL with 0.12 N HCl and set aside as the control. The remaining six test tubes were placed in a 150°C oven to evaporate free water. After one hour of drying, one test tube was removed from the oven, diluted to 10 mL with 0.12N HCl, and labeled as the 150°C data point control sample. The remaining five test tubes were placed in a muffle furnace at 200°C. After thirty minutes in the furnace, one test tube was removed from the furnace, diluted, labeled, and set aside for analysis. The temperature of the furnace was increased in 50°C increments after the removal of each test tube. This step was repeated until all five test tubes were removed from the furnace.

RESULTS AND DISCUSSION

Results from the decomposition test are shown in Figure 7.1. At 250°C, 96% of the perchlorate was decomposed based on a starting concentration of 5,600 mg/L and a final concentration of 220 mg/L. Below this temperature, only a small fraction of perchlorate was decomposed. Although temperature had to be increased to 400°C to obtain complete decomposition in this experiment, temperatures in the range of 250 to 300°C may be adequate for a full-scale application for two reasons. First, complete reduction of perchlorate is not required in the ferric chloride regeneration system. Near complete resin regeneration can be accomplished with significant concentrations of perchlorate present in the regenerant because the resin has much greater affinity for the tetrachloroferrate ion ($FeCl_4^-$) compared to the perchlorate ion. Second, full-scale thermal decomposition systems may provide faster and more efficient heat transfer resulting in higher degradation rates at a given temperature. During this process, nitrate in the surrogate (500 mg/L) was completely reduced at a much lower temperature of 150°C.

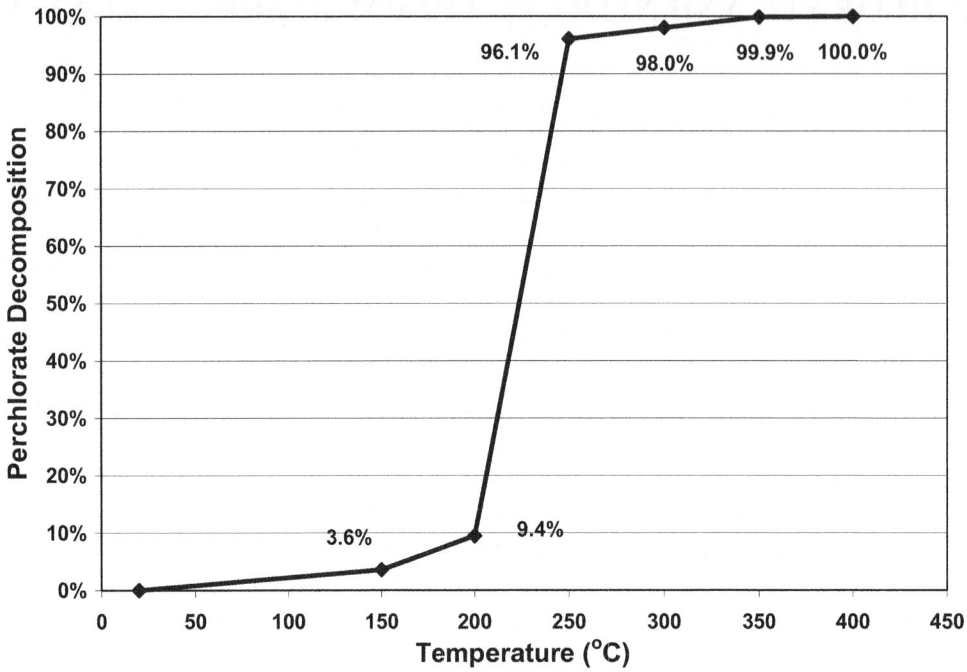

Figure 7.1 Thermal decomposition temperature results

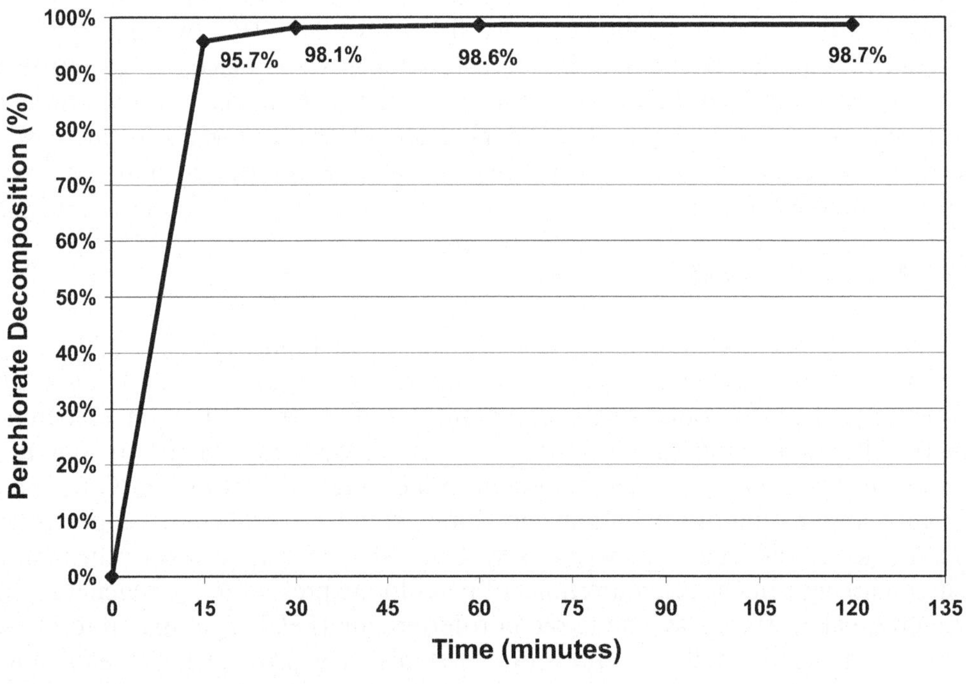

Figure 7.2 Thermal decomposition of perchlorate at 250°C

Based on the results from initial decomposition tests, 250°C was selected as the test temperature for evaluating decomposition rate. The experimental setup and procedure was similar to the previous decomposition tests. Again, 0.5 mL of the surrogate solution was added each of six test tubes, one of which was diluted to 10 mL with of 0.12N HCl and set aside as the control. The remaining five test tubes were placed in a 150°C oven to evaporate free water. After one hour, the test tubes were removed from the oven and placed in a muffle furnace set at 250°C. Test tubes were removed from the furnace at 15, 30, 60, and 120-minute intervals, cooled, diluted, and labeled in preparation for analysis. Results from the rate test are shown in Figure 7.2. The results are consistent with the temperature test in that the majority of the perchlorate is rapidly reduced at 250°C. Although complete reduction was not obtained at 250°C after 2 hours, complete reduction is not necessary as mentioned previously.

SUMMARY AND RECOMMENDATIONS

Based on the testing conducted on ferric chloride residuals, two technologies were identified that could be implemented to achieve perchlorate destruction and partial recovery and reuse of components. These commercially available technologies are spray roasting and spray drying.

Spray roasting has been used for many years to process spent pickle liquor from steel pickling processes (hydrochloric acid treating to remove surface iron oxide). Pickle liquor is essentially a dilute version of spent ferric chloride regenerating solution. Spray roasting is conducted at relatively high temperatures (600°C to 900°C). The most valuable product of spray roasting is iron oxides used as pigments in coatings and plastics. In the spray-roasting environment, perchlorate is readily destroyed and residual HCl is evaporated. In addition to the residual HCl, HCl is also generated from pyrohydrolysis of ferric chloride according to the reaction:

$$FeCl_3 + 1.5\,H_2O \rightarrow 0.5\,Fe_2O_3 + 3\,HCl$$

Therefore, a net surplus of HCl can be recovered from this process by HCl absorbers. HCl is recovered as an azeotrope at a concentration up to 18% and can be used to prepare regenerating solution. Excess HCl can be used for the resin rinse process and could also be used to convert recovered iron oxide back to ferric chloride for reused in the regenerating solution. Based on projected spent regenerating solution composition (using concentrated HCl), products of spray roasting one liter of solution are approximately 1.6 L of 18% HCl and 90 grams of iron oxide. If 18% HCl is used in the preparation of the regenerating solution, lower yields of HCl will be obtained. Spray roasting is an energy intensive process requiring about 80 watts of electricity and up to 800 Kcal of heat, from natural gas, per liter of solution treated.

There are operational issues that must be addressed to use spray roasting technology for this application. Most spray roasting plants operating today are much larger than would be required for handling projected effluent from the ferric chloride regeneration processes. Smaller spray roasting plants would be less efficient. Existing plants could be used to dispose of waste regenerating solution. However, impurities, including sulfate and other minerals, could affect the quality of iron oxide and limit the number of facilities willing to accept this effluent. Treatability tests would have to be conducted to determine the impact to iron oxide quality or plant maintenance necessary to return to standard operation.

Spray drying may also be used to destroy perchlorate and recover HCl and iron compounds. Spray drying is conducted at much lower temperatures and exhibits a significant temperature gradient. Inlet temperature up to 550°C can be obtained but outlet temperatures are usually in the range of 100 to 200°C. Outlet temperature can be controlled by the liquid feed rate. Feeding at rates lower than typical design conditions may enable higher outlet temperatures to be maintained and high perchlorate destruction rates to be achieved. Pilot studies are recommended to determine perchlorate reduction performance, the effect of co-contaminants, and the form and quality of the iron product. Unlike spray roasting equipment, pilot-scale spray drying equipment is commercially available. Based on the destruction tests conducted as part of this study, spray drying of spent ferric chloride regenerating solution should generate temperatures high enough to effect nearly complete perchlorate decomposition. All residual HCl is recoverable, but additional HCl from pyrohydrolysis of $FeCl_3$, may not be generated.

For this process to be implemented, it is recommended that alternative regeneration technologies using $FeCl_3/HCl$ for selective resin be considered and further evaluated. Such follow-up work is warranted because of the potential for decreased volume of generated wastes as well as recovery and reuse of regenerant solution components.

CHAPTER 8
PERCHLORATE AND NITRATE TREATMENT AT A WWTP

INTRODUCTION AND OBJECTIVES

Currently, options for handling/treating concentrated perchlorate- or nitrate-laden liquid waste streams include discharge to a brine line or reduction using dedicated brine treatment systems. One novel treatment approach is to discharge the concentrated waste stream to a sewer and biologically reduce the perchlorate and/or nitrate at the receiving wastewater treatment plant (WWTP). Unique aspects of this approach are: 1) the bioreactor at the WWTP is acclimated with indigenous organisms (i.e., no exogenous microbial enrichment or seeding would be required), and 2) background organics serve as the substrate for oxidant reduction (i.e., no exogenous substrate is spiked to the system). The overall objective of this study is thus to investigate the efficacy of using biological processes to treat perchlorate- and nitrate-laden concentrates at a WWTP.

The first part of this study investigated the feasibility of using wastewater oxidation ditches to treat perchlorate residuals. Since biological perchlorate reduction requires anaerobic conditions, renovation of existing oxidation ditches would be necessary to allow for the development of both aerobic and anaerobic zones. One option would be to provide point aeration at certain locations in a given oxidation ditch. Ideally, the aerators would be positioned some distance downstream of the feed line so that the oxidation of ammonia, which forms nitrate and thus inhibits perchlorate reduction, occurs after perchlorate has been removed. This would promote the development of anaerobic conditions at the front end of the oxidation ditch, where perchlorate reduction can occur. Farther downstream, aerated zones would provide the oxygen necessary for biochemical oxygen demand (BOD) removal (USEPA 1993).

The second part of this study investigated the feasibility of using a fixed-film reactor to treat perchlorate residuals. In contrast to suspended growth in a mixed tank, a fixed-film reactor would focus the growth of microorganisms in biofilms that form on a support medium such as sand, plastic, or GAC. The biogrowth support media can either be fixed or fluidized. A fluidized-bed requires high feed pumping rates, but does not have to be taken off-line to clean the bed; a fixed-bed uses lower feed pumping rates, but requires periodic backwashing to control headloss build-up. At the full-scale, a fixed-film reactor would be placed just downstream of the headworks (i.e., screens, grit channels, comminutors, etc.).

There are several reasons why a fixed-bed biological reactor might be conducive to achieving rapid perchlorate degradation:

- The perchlorate-reducing metabolic activity of a biofilm can be more stable than that of suspended cultures. Research has shown that when suspended perchlorate-reducing bacteria are exposed to dissolved oxygen (DO), their perchlorate-reducing metabolic activity recovers slowly when anaerobic conditions are established. Alternatively, perchlorate degradation kinetics in fixed-film processes are not significantly impacted by transient DO or nitrate exposure.
- A gradient of redox potential can be developed across the depth of the bed. This allows for the development of semi-distinct DO-, nitrate-, and perchlorate-reducing

zones within the bed, which forces bacteria to use a specific metabolic activity instead of continuously altering metabolic states.

- A redox potential gradient can be established across the depth of a given biofilm, which would permit perchlorate reduction even if the bulk solution is not anaerobic.
- A fixed-bed process can be placed just downstream of the headworks, which will maximize the availability of readily degradable organic carbon required for perchlorate degradation and prevent the introduction of dissolved oxygen (due to aeration) and nitrate (due to nitrification).

For this project, a bench-scale, GAC-based, fixed-bed biological reactor was tested. A bench-scale study was conducted to determine the viability of removing perchlorate residuals from wastewater using biological treatment. Mixed liquor suspended solids (MLSS) and wastewater from Magna Water Company's Wastewater Treatment Plant (WWTP) (Magna, Utah) were used to establish perchlorate-reducing biological activity and to provide substrate for perchlorate-reducing organisms, respectively. The objectives of this study were to: (1) determine whether or not bacteria indigenous to Magna Water Company's WWTP could reduce perchlorate; (2) determine whether or not the organic matter present in Magna wastewater could serve as substrate for biological perchlorate reduction; (3) determine the contact time required to achieve perchlorate removal to below detection in a simulated, bench-scale anoxidation ditch. (4) compare the perchlorate reduction kinetics of a fixed-film process to those observed for the suspended growth process. (5) determine the minimum empty bed contact time (EBCT) required to achieve perchlorate removal to below detection using a fixed-film bioreactor.

MATERIALS AND METHODS

The bench-scale studies were conducted at the Carollo Research Laboratory in Boise, Idaho. A batch test apparatus consisting of a 10-gallon Teflon® tank with a headspace-free cover (to prevent the diffusion of oxygen into the influent) was used to simulate a full-scale oxidation ditch. Wastewater and MLSS were collected from Magna's oxidation ditch influent line and return activated sludge line (Figure 8.1), respectively, and shipped overnight at ambient temperature (~22°C) to Boise in 5-gallon carboys. The water was stored in a refrigerator until the start of each experiment while the MLSS was stored at room temperature. Ten gallons of wastewater were poured into the batch reactor and spiked with ~1 mg/L of perchlorate. MLSS was then added to the batch reactor to reach a final concentration of 2,000 mg/L, which matches the average full-scale MLSS concentration in the oxidation ditches. The reactor was placed on a stir plate that mixed the wastewater throughout the testing period.

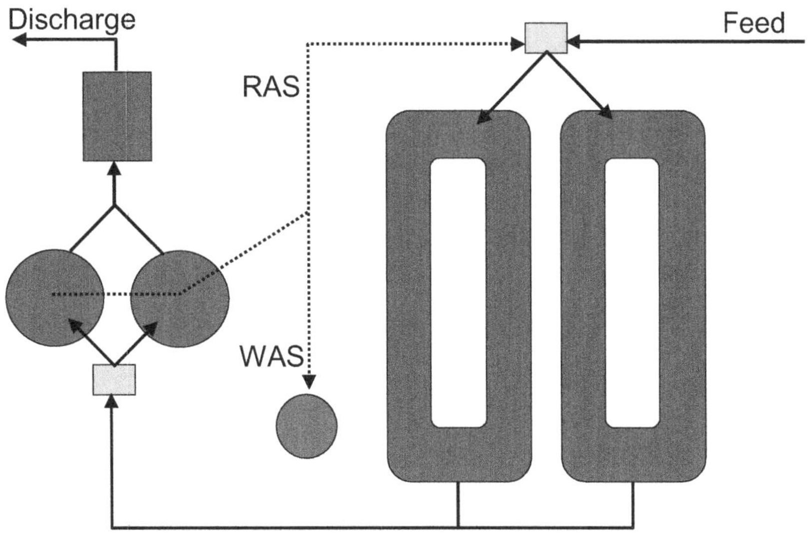

Figure 8.1 Magna WWTP schematic (WAS: waste activated sludge; RAS: return activated sludge)

DO and nitrate concentrations were below detection in the raw wastewater. Perchlorate samples were taken each hour for the first 3 hours and then approximately every 24 hours thereafter up to 67 total hours of incubation time. Only three perchlorate samples were initially sent for analysis so that the degradation curve could first be located before other samples were analyzed to resolve the curve. Once the results of the first experiment were available, the experiment was repeated using a new set of wastewater and MLSS samples and a lower initial perchlorate concentration (~0.6 mg/L).

DO, pH, nitrate, TOC, and DOC concentrations were measured in the wastewater immediately after sampling at Magna's WWTP and upon arrival in Boise. This served to identify any water quality differences between bench- and full-scale, which may impact perchlorate removal performance. Perchlorate and nitrate concentrations were measured using EPA Method 314.0 and 353.2, respectively, by a commercial lab (SVL Analytical, Kellogg, Idaho). The method detection limits for perchlorate and nitrate in the wastewater matrix were 0.008 mg/L and 0.05 mg/L, respectively. Samples were stored in a refrigerator (~5°C) until shipment to the analytical lab. Samples were shipped in a cooler filled with ice packs (~2°C). TOC and DOC were measured by UV-Persulfate oxidation using a TOC analyzer (SM 5310C). pH was measured using a portable HACH probe.

For the fixed-film experiments, an activated carbon filter was constructed using Calgon F-400 GAC, a bituminous-based, pellet-shaped carbon. The carbon was placed in a 2-inch (5.08-cm) inner diameter glass pipe (Ace Glass, Vineland, N.J.), which was capped with threaded Teflon® endcaps. Raw wastewater was pumped from a 10-gallon Teflon® influent tank to the bioreactor through Norprene® tubing (Masterflex, Cole Parmer, Vernon Hills, Ill.) using a peristaltic pump (Masterflex, Cole Parmer, Vernon Hills, Ill.). A floating cover was placed inside the reservoir to minimize oxygen diffusion into the influent. For the first portion of testing, a perchlorate stock solution was spiked to the influent reservoir to reach a final concentration of 1 mg/L. However, perchlorate degradation in the influent made it difficult to maintain feed perchlorate concentrations near 1 mg/L. Therefore, a syringe pump (kd Scientific, CP Instrument

Company Ltd., United Kingdom) was added to the system to dose perchlorate to the feed just prior to entering the fixed-bed. A 0.2 μm filter cap was attached to the end of the each syringe to prevent biogrowth from impacting the perchlorate stock solution. For the 60- and 30-minute EBCT experiments, the carbon bed depth was maintained at 7 inches (17.78 cm). Due to limitations on the availability of wastewater, a carbon bed depth of 1.2 inches (3.05 cm) was maintained for the 5- and 10-minute EBCT experiments. No inoculum was used as the biological activity was developed by contacting the bed with perchlorate-spiked wastewater. A photo of the fixed-bed experimental apparatus and a close-up photo of the syringe pump connection are shown in Figures 8.2 and 8.3, respectively.

RESULTS AND DISCUSSION

Wastewater Quality Parameters

The Magna WWTP has an average daily flow of 3.3 mgd and a peak daily flow of approximately 7 mgd. Average feed water parameters are as follows:

- NO_3^--N (mg/L) = 0.9 mg/L.
- TKN (comprised mostly of NH_3) (mg/L) = 45.
- BOD (mg/L) = range: 148-224; 2002 average = 185.
- pH = 7.3.

The WWTP also receives an intermittent perchlorate discharge from a local solid-rocket manufacturing company, which can result in average feed perchlorate concentration near 1 mg/L. Table 8.1 compares wastewater quality data before and after overnight shipment from Magna to Boise. While DO and pH remained constant, approximately 4 mg/L of nitrate degraded to below detection during shipment. TOC and DOC appeared to increase between sampling at the WWTP and arrival in Boise. Therefore, a second set of TOC/DOC samples were taken. These samples, taken two months after the first set, showed that 7 to 8 mg/L of organic carbon degraded between sampling at the WWTP and arrival in Boise.

Table 8.1
Wastewater quality preservation (data from Experiment 2)

Parameter	After sampling at WWTP	Upon arrival in Boise
TOC (mg/L)	12/41* (0.5/1.0)†	60.3/33.6 (0.19/1.00)
DOC (mg/L)	12/28 (0.5/1.0)	43.7/20.3 (0.19/1.00)
NO_3^- (mg/L as N)	0.9 (0.1/0.1)	ND (0.008/0.050 mg/L as NO_3^-)
DO (mg/L)	ND (0.5/0.5)	ND (0.1/0.1)
pH	7.2	7.1
Perchlorate (μg/L)	ND (2.0/4.0)	ND (0.0013/0.004 mg/L))

* Second value is from the second TOC/DOC test
† (X/Y) = (Method detection limit/Minimum reporting limit)
ND: non-detect level

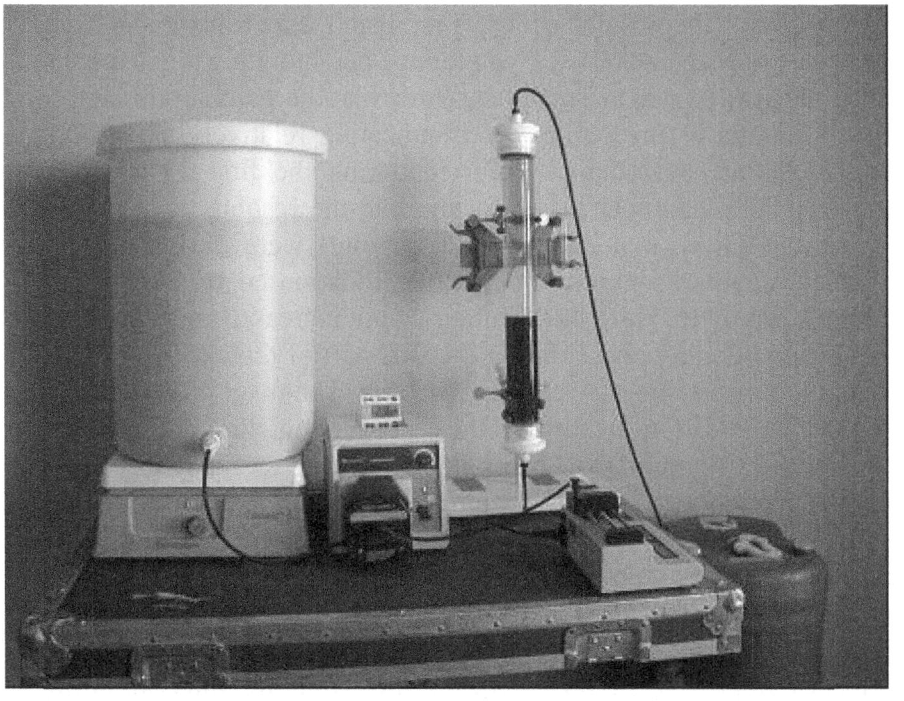

Figure 8.2 Fixed-bed experimental apparatus for perchlorate treatment at a WWTP

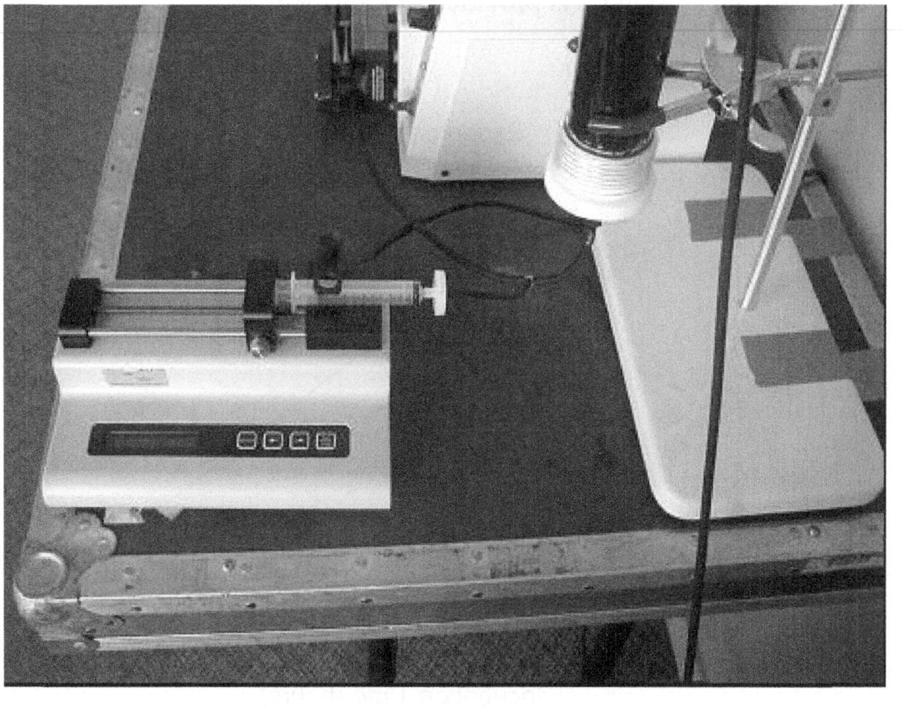

Figure 8.3 Syringe pump connection for WWTP perchlorate treatment test set-up

Anoxidation Ditch Results

Figure 8.4 shows the results of the first batch experiment. The data indicated that organisms indigenous to Magna's WWTP can reduce perchlorate while using the natural organic matter present in the wastewater as substrate. However, the perchlorate concentration was still 0.22 mg/L after 67 hours of incubation time, indicating relatively slow degradation kinetics. Since minimal perchlorate degradation occurred during the first 33 hours of incubation, no additional samples, which had been taken to improve the resolution of the degradation curve, were sent for analysis. Given that only three data points were measured, it is not possible to identify exact kinetic regimes within the overall degradation curve. However, it is evident that perchlorate degradation rates were low initially and increased substantially sometime after 33 hours of incubation. The lag phase during the first part of the experiment may have been due to a low concentration of perchlorate-reducing bacteria. The mass of perchlorate reducers would have increased as degradation occurred, thus accounting for the progressively increased degradation rates. The slower initial perchlorate degradation rate may also have been due to the fact that all known perchlorate-reducing bacteria are facultative aerobes or microaerophile (i.e., they respire aerobically when DO is present). To achieve anaerobic respiration using perchlorate as an electron acceptor, these bacteria must first synthesize the requisite perchlorate-reducing enzymes, thus causing a lag phase.

When the experiment was repeated with a fresh batch of wastewater, perchlorate degradation kinetics improved slightly. Due to the sampling interval during Experiment 2 (Figure 8.3), the exact time at which perchlorate reached ND cannot be determined. However, extrapolation of the data suggests that perchlorate would have been ND by approximately 45 hours, as shown by the dashed line in Figure 8.5. A comparison of perchlorate degradation kinetics between Experiments 1 and 2 is provided in Table 8.2.

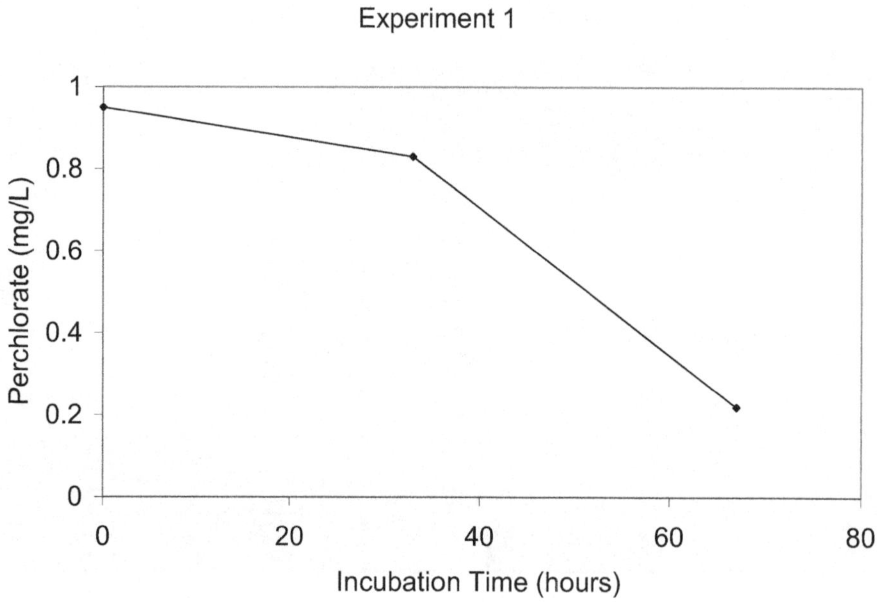

Figure 8.4 Perchlorate concentrations as a function of incubation time (1st batch test)

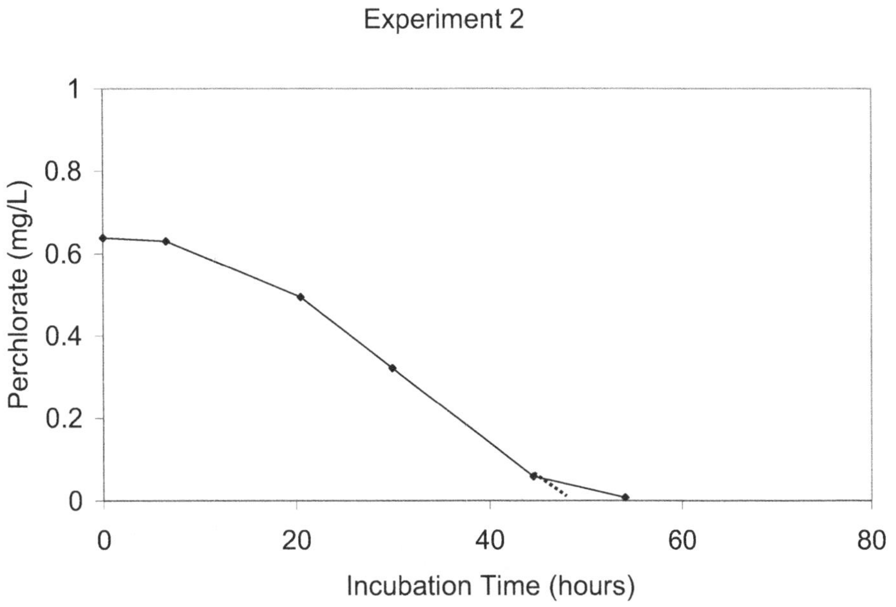

Figure 8.5 Perchlorate concentrations as a function of incubation time (2nd batch test)

Table 8.2
Rate of perchlorate reduction*

Experiment 1		Experiment 2	
Time interval (starting hr-ending hr)	Reduction rate (mg/L·hr)	Time interval (starting hr-ending hr)	Reduction rate (mg/L·hr)
0-33	0.004	0-6.5	0.001
		6.5-20.5	0.010
33-67	0.018	20.5-30	0.018
		30-44.5	0.018

* Kinetic rates were calculated for relative comparison purposes only.

Though a lag phase was observed during Experiment 2, it was substantially shorter than that observed during Experiment 1. One explanation for this is that while Experiment 2 started within 24 hours of the second shipment of MLSS and wastewater, Experiment 1 was not initiated until 5 days after the receipt of the first MLSS and wastewater shipment. With no DO, nitrate, or perchlorate present, fermenting bacteria might have thrived while bacteria capable of reducing perchlorate decayed. Thus, it is possible that fewer active perchlorate-reducing bacteria were present in the initial batch solution during Experiment 1. Table 8.2 also shows that post-lag phase perchlorate removal kinetics were similar between Experiments 1 and 2 at 0.018 mg/L·hr.

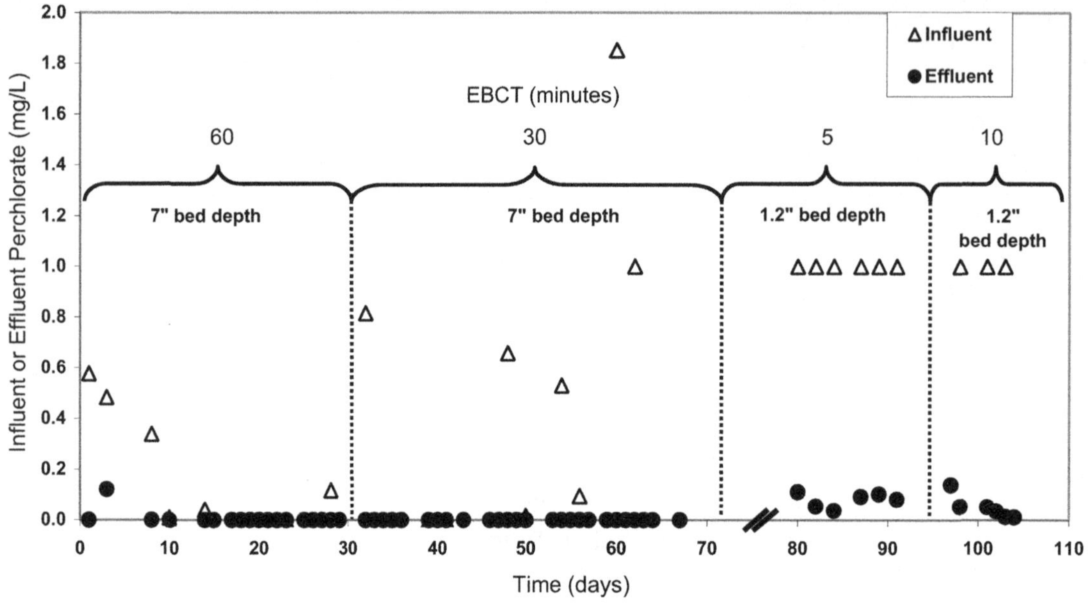

Figure 8.6 Fixed-bed biological perchlorate removal performance

Fixed-Film Reactor Results

Perchlorate removal performance for the fixed-bed biological reactor is shown in Figure 8.6.The figure is divided into four testing regimes, reflecting the four EBCTs tested: 60, 30, 5, and 10 minutes. The data reveal several phenomena:

The 1 mg/L influent perchlorate concentration was difficult to maintain. Since new influent was made approximately every 3 days, there was sufficient time for perchlorate to degrade in the feed tank. Several corrective measures were implemented in an attempt to maintain a 1 mg/L feed perchlorate concentration, including doubling the initial feed concentration to 2 mg/L (Day 30), removal of biomass from the influent reservoir (Day 39), filtering (0.45 μm) and aerating samples (Day 43), and diluting samples (Day 60). Though these measures were somewhat effective, feed perchlorate concentrations were still typically less than 1 mg/L. This problem was solved after the 30-minute EBCT experiment when a syringe pump was added to the system for perchlorate dosing. Sterile syringes were used and the perchlorate stock solution was drawn into each syringe through a 0.2 μm filter, thus sterilizing the solution.

Perchlorate was initially removed to below detection (4 μg/L), which was likely the result of adsorption onto the GAC. Perchlorate breakthrough of 25% was observed on Day 3, reflecting the loss of adsorption capacity as perchlorate saturated the carbon. Perchlorate removal to below detection was again observed on Day 8, suggesting that perchlorate-reducing biological activity had developed in the bed. The fixed-bed achieved and sustained perchlorate removal to below detection at 60- and 30-minute EBCTs. Appreciable influent perchlorate concentrations ranged from 16 μg/L to 1.85 mg/L as shown in Figure 8.6.

At a 5-minute EBCT, 89 to 96% perchlorate removal was observed. Given this performance, it was expected that doubling the EBCT to 10 minutes would result in consistent perchlorate removal to below detection. However, only 86-99 percent perchlorate removal was achieved at a 10-minute EBCT. These data suggest that something other than contact time was limiting the system. One possibility is that the concentration of perchlorate-reducing biomass in the bed was inadequate to achieve perchlorate removal to below detection. The carbon bed depth was decreased to 1.2 inches (3.05 cm) to allow for the evaluation of the 5- and 10-minute EBCTs without consuming substantial volumes of feed water. At a bed depth of 1.2 inches, the volume of carbon was only 3.8 in^3 (62.3 cm^3), a 5.78-fold volume reduction relative to the 7-inch bed-depth scenario. Since there is a finite concentration of biomass that can be supported per unit volume (or surface area) of GAC, spreading the contact time over a larger volume of biologically active GAC would likely increase the ratio of biomass to contact time, thus improving perchlorate removal performance.

Treatment Process Considerations

There are numerous factors that must be considered when implementing a full-scale biological process to treat perchlorate at a WWTP. One of such factors is a treatment progress concept. Dedicated perchlorate treatment processes may not need to be considered if perchlorate discharges to a sewer system are relatively low while the treated water volumes at the WWTP are substantial. In other words, dilution may drive perchlorate concentrations to below the limits of detection at the WWTP.

Though the bench-scale fixed-film experimental results were more promising than the anoxidation ditch simulation results, suspended growth processes should not be eliminated immediately from consideration. For example, any full-scale WWTP process that includes a dedicated anaerobic zone (e.g., biological phosphorous removal processes) may be particularly well-suited to treat perchlorate, and may not require any modifications to the existing treatment facility. In addition, other fixed-film configurations, such as a fluidized-bed reactor, were not tested and may need to be considered for full-scale treatment. It is also important to note that the intermittent discharge of perchlorate to the sewer may impact the perchlorate-reducing metabolic activity of any bioreactor at the WWTP. This work did not investigate how intermittent perchlorate concentrations might affect perchlorate removal kinetics. Further testing is required to isolate this impact.

Design Considerations

In general, perchlorate- or nitrate-laden residuals that are treated at a WWTP should be completely destroyed in the process. That is, perchlorate and nitrate should be converted to innocuous chloride, water, and nitrogen gas, effectively eliminating the need for any additional treatment or handling of the initial residual stream.

Further research is necessary to determine the optimal choice for the biogrowth support medium in a fixed-film reactor (i.e., a different type of GAC or a different type of media altogether such as sand or plastic). Minimum required EBCT will not likely be the same among different biogrowth support media. However, it is possible that a perchlorate degradation rate can be correlated to the specific area (i.e., surface area per volume) of media, which may be roughly constant among different media. Additional testing is also required to establish an effective backwashing protocol and to examine the impact of particulate BOD removal by a fixed-bed filter on the operation of a downstream activated sludge process.

Siting constraints of implementing a dedicated, fixed-film perchlorate treatment process should be evaluated on a case-by-case basis. Results from this work suggest that the required EBCT for complete perchlorate removal is low, thus requiring a relatively small footprint to house the fixed-film reactor. It is also anticipated that a fixed-bed process might best be placed just upstream of secondary treatment to maximize the availability of substrate and to minimize DO and nitrate inhibition. If there is no primary sedimentation, special consideration should be given to the presence of large suspended matter that could clog the fixed-bed. Additional screening may be required upstream of the fixed-bed to address this concern.

Other Issues

The practice of discharging perchlorate to the sewer may be particularly applicable to utilities that operate and manage both the water and wastewater treatment plants. In situations where the utilities operate independently, complicated interagency agreements may need to be reached that allow for perchlorate residuals to be discharged to the sewer without violating discharge limits. Responsibility for covering capital and O&M costs associated with treating perchlorate residuals at a WWTP would need to be worked out between the wastewater utility and the party responsible for discharging perchlorate to the sewer.

An alternative to the approach of discharging perchlorate/nitrate-laden concentrates to the sewer and treating the entire WWTP flow would be to construct a dedicated bioreactor to treat the concentrates after they have been blended with a portion of the total municipal wastewater flow, which has been scalped from a local sewer line. With this approach, the bioreactor would be constructed near or at the drinking water treatment facility, and the concentrate or brine stream from the ion exchange, reverse osmosis, or electrodialysis reversal process would be blended with scalped wastewater prior to treatment in the bioreactor. Blending the oxic waste stream with municipal wastewater will decrease the bulk DO and total dissolved solids concentration, which should improve the kinetics of perchlorate and nitrate degradation. Background organics present in the municipal wastewater would still serve as the substrate for biological nitrate and perchlorate reduction and the bioreactor would still be acclimated with organisms indigenous to the local wastewater. Using this approach, the total flow being treated in the bioreactor should be substantially smaller than the flow applied to the bioreactor if it were incorporated into the existing WWTP. Thus, significant capital cost savings might be realized using this approach.

SUMMARY AND RECOMENDATIONS

The bench-scale oxidation ditch simulation experiments demonstrated that bacteria indigenous to Magna Water Company's WWTP can reduce perchlorate using natural organic matter as substrate. However, hydraulic residence times of at least 2 days were required to achieve perchlorate degradation to below detection. Therefore, although the (an)oxidation ditch process did work, perchlorate degradation kinetics must be enhanced before full-scale implementation of this or a similar process can be considered. This might be done through the addition of an exogenous electron donor such as methanol. However, it appears as though there is sufficient natural organic carbon present to achieve perchlorate reduction, suggesting that methanol addition will not significantly improve perchlorate degradation kinetics.

Seeded only with microorganisms indigenous to the raw wastewater, the fixed-bed biological reactor demonstrated efficient perchlorate removal while using only background organics as substrate. Perchlorate removal to below detection was achieved and sustained at EBCTs as low as 30 minutes, a considerably shorter period than was required to achieve perchlorate removal to below detection in the simulated anoxidation ditch system (48 hours). Experiments at 5- and 10-minute EBCTs also demonstrated effective perchlorate removal, in spite of the small carbon bed volume used for these experiments. Additional testing is required to determine whether or not perchlorate removal to below detection can be achieved and sustained at EBCTs shorter than 30 minutes.

The results of this work indicate that perchlorate-laden concentrates combined with municipal wastewater may effectively be treated using a fixed-bed biological process. However, it is recommended that additional research be performed to resolve the remaining unknowns related to the design, start-up, and operation of such a process. Some of these unknowns include media selection, backwash protocol, impact of intermittent perchlorate exposure on degradation kinetics, and temperature effects. It is also recommended that studies be performed to investigate the efficacy of using a bioreactor to treat concentrate streams that have been blended with scalped municipal wastewater, an approach that might be more cost-effective than treating the entire wastewater flow.

CHAPTER 9
UTILITY GUIDELINES

An integrated approach must be used when defining/designing new treatment processes for arsenic, perchlorate, and nitrate to fully understand and capture all potential issues associated with wastes generated during each one of them. The potential type of residuals from the selected treatment approaches identified in Chapter 2 of this report can significantly increase the overall cost to produce water. This section of the report presents some of the key issues and guidelines to be considered by the utilities when selecting processes for arsenic, perchlorate, and nitrate treatment. The purpose of this section is to provide more in-depth understanding of the implementation issues at the treatment facilities. Each utility, however, must define project-specific goals and objectives and use this document only as a guidance document. The following categories are important in developing fully integrated residual management approach:

- Types of residuals produced (e.g. liquid, solid, high brine)
- Applicable discharge, disposal, and transportation regulations (types and limits).
- Proximity to regeneration, disposal and discharge facilities.
- Capital and O&M cost for residual treatment.

INTEGRATION OF RESIDUALS MANAGEMENT IN TREATMENT SELECTION

This section presents schematics of conceptual treatment trains including the treatment process (solid lines), the wastes generated (dashed lines), the possible waste treatment required (double dashed lines), and disposal methods. Identifying the residual stream is the first step in assessing more complete cost estimates, and these schematics can assist the utilities in identifying the potential residual issues prior to implementing the treatment processes.

Integrated Process Trains for Arsenic

Unlike perchlorate or nitrate treatment processes for which the contaminants can be degraded given the right conditions, all arsenic treatment technologies rely on phase transfer methods to remove arsenic from the bulk water and concentrate it in a smaller volume to be discharged or disposed of. One such technology is single-use media adsorbent such as GFH, GFO, single-use AA, iron-incorporated resin, and TiO_2 media. Similar to single-use ion exchange for perchlorate application, these single-use media for arsenic are replaced every few months. The spent media are disposed of in various class of landfills depending on leaching test results. Backwash water from this process, as shown in Figure 9.1, contains low levels of arsenic. Depending on the location of an installation and the actual arsenic level in the backwash, this waste stream may be discharged to a sewer, or further treatment or recycle of this stream can be done to minimize the discharge of arsenic waste as discussed in Chapter 3 of this report.

Certain types of media, such as AA may be regenerated on-site for continued use. The regeneration typically requires chemicals such as sodium hydroxide and sulfuric acid for conditioning the media, and the spent chemical solution must be treated, neutralized and disposed of properly. During the chemical regeneration with base and acid, deterioration of media occurs and the arsenic sorption capacity diminishes in the subsequent cycle. The conceptual schematic of this process is illustrated in Figure 9.2.

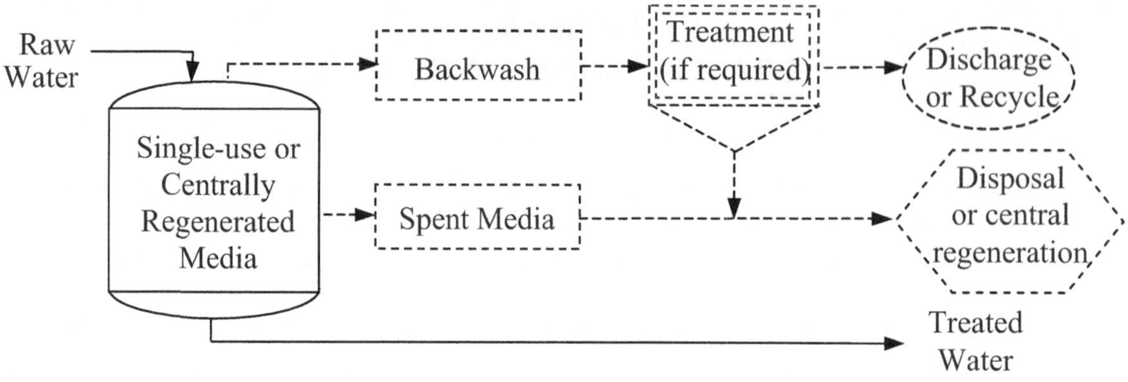

Figure 9.1 Schematic of single-use media adsorbent for arsenic treatment

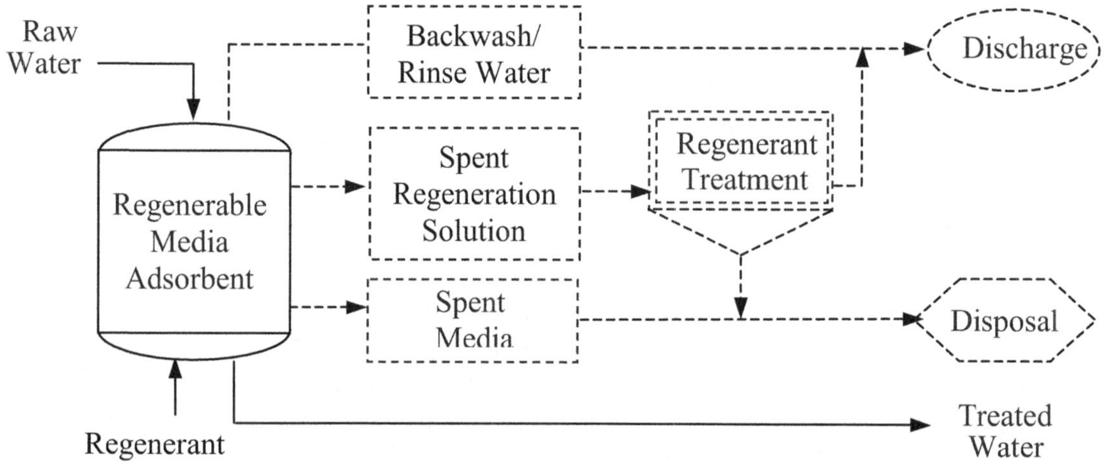

Figure 9.2 Schematic of regenerable media adsorbent for arsenic treatment

The regenerable ion exchange system for arsenic treatment is similar in design to that for nitrate or perchlorate (Figure 9.3). The only difference in the arsenic treatment application is in handling of the spent brine. Unlike the perchlorate or nitrate applications for which contaminants present in the brine can be destroyed biologically or thermally prior to discharge, arsenic-laden brine must be settled or solidified so that arsenic can be disposed of in a landfill. One approach is to use hydrogel sorbent as described in Chapter 5 of this report.

Since desalting membrane processes (RO, NF, and EDR) produce a large quantity of waste (membrane concentrate), this may decrease the concentration of arsenic in the reject stream is less concentrated than the spent ion exchange brine. However, if the arsenic concentration in the reject stream is less than the local discharge limit, the waste can be discharged without further treatment. If the level exceeds the local limit, then further treatment of the waste will be required, which will significantly increase the treatment cost given the large volume of waste that must be treated. As previously mentioned, the membrane cleaning waste may be discharged to sewer after pH adjustment, as illustrated in Figure 9.4.

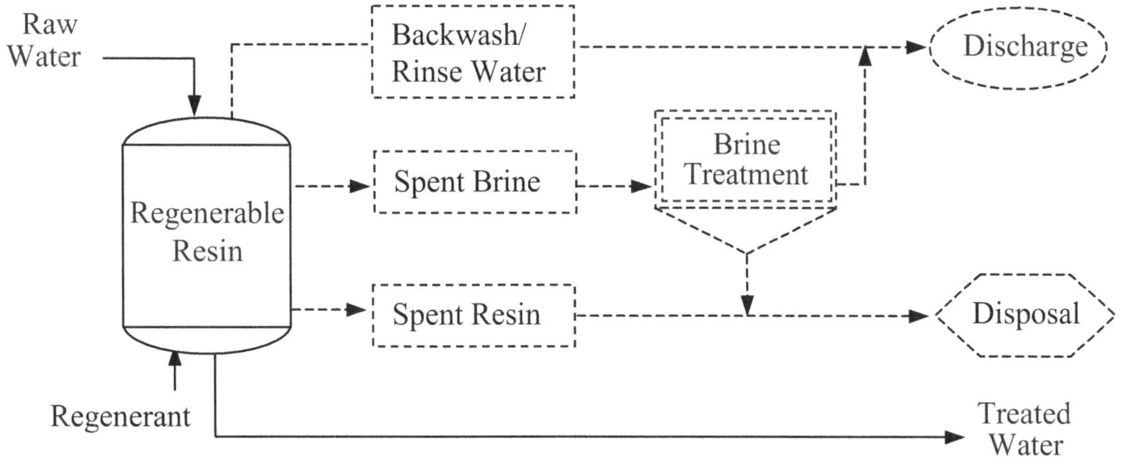

Figure 9.3 Schematic of regenerable ion exchange for arsenic treatment

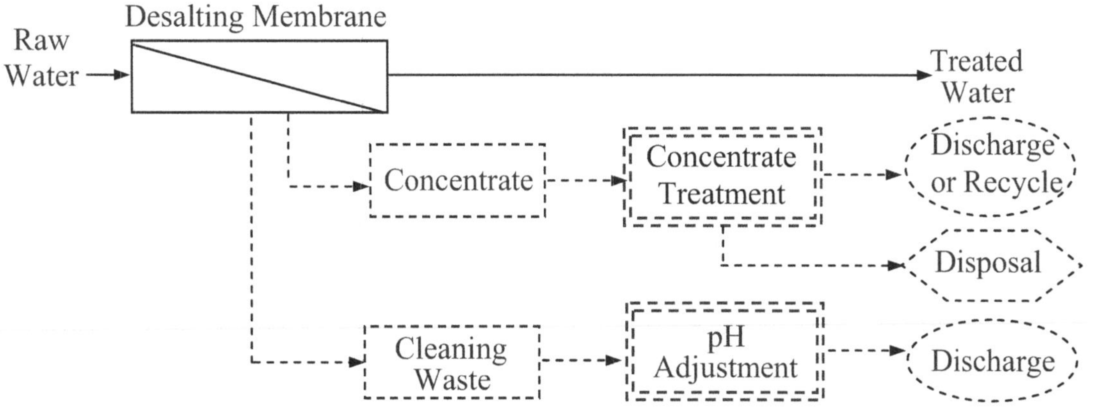

Figure 9.4 Schematic of desalting membrane process for arsenic treatment

Another type of arsenic treatment is based on proven coagulation filtration technology. Typically, coagulant (e.g., ferric) and oxidant (e.g., chlorine) are added, and arsenic is complexed and filtered with a barrier such as a media filter, a greensand filter, or a microfilter (Figure 9.5). The backwash water containing arsenic must be settled, and the settled sludge is dewatered and may be disposed of with or without treatment depending on the leaching test result. The supernatant water from the settling may be discharged or recycled. The media does not need to be replaced as it is used for filtration, not adsorption. The backwash frequency and the volume of residual produced greatly depend on water quality and operational requirements.

For large systems, conventional treatment or direct filtration with coagulant (alum or ferric) is capable to treat large volumes of water for arsenic removal. Depending on the treatment goal, the coagulant dose may be high. Figure 9.6 shows a schematic for a conventional system with settling, and for this case arsenic is concentrated in the settled sludge. This sludge must be dewatered and disposed of in a landfill, again depending on leaching test results. The filter backwash does not typically contain arsenic and can be discharged to the sewer.

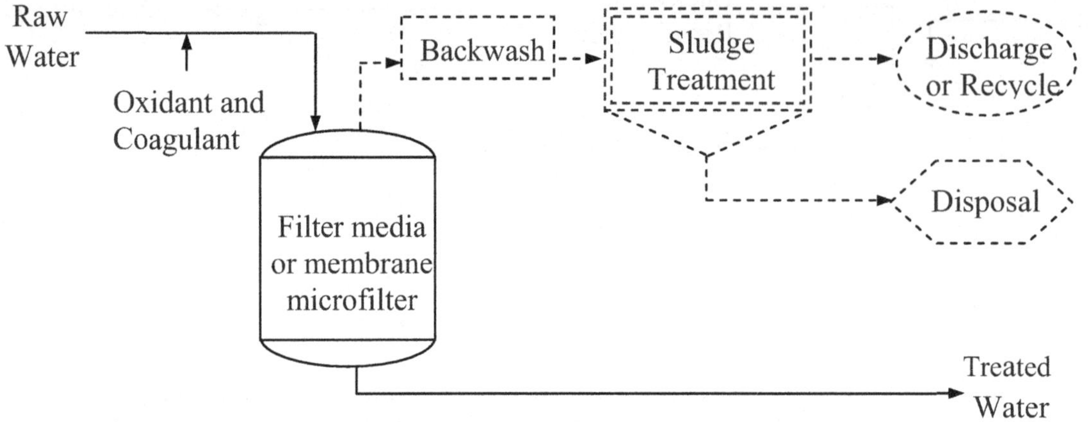

Figure 9.5 Schematic of coagulation filtration for arsenic treatment

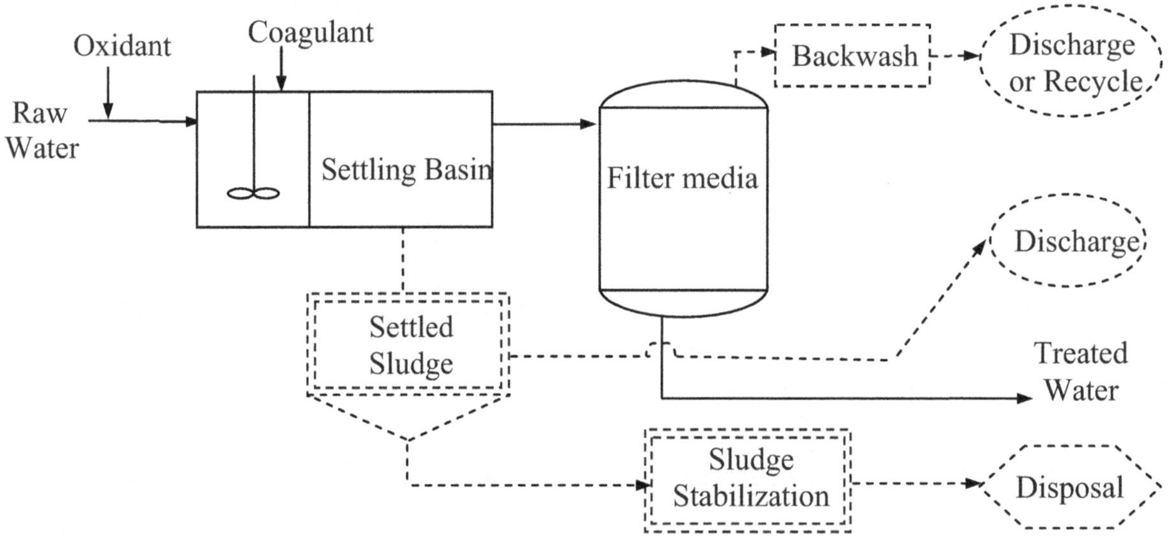

Figure 9.6 Schematic of conventional treatment for arsenic removal

Integrated Treatment Trains for Perchlorate and Nitrate

The biological treatment process, such as the fluidized bed system shown in Figure 9.7, is a destruction technology in which nitrate or perchlorate are destroyed by microorganisms attached in the bed. Thus the only waste produced is excess biomass, which can be discharged to a sewer. Another type of the biological process is a fixed bed system (Figure 9.8). Similar to the fluidized bed, fixed bed systems produce only excess biomass waste as both nitrate and perchlorate are destroyed during the treatment. These biological processes require addition of a carbon-based electron donor for microbial processes as indicated in the schematics. Typical end products in the treated water are chloride, nitrogen, carbon dioxide, and water.

The membrane biofilm reactor is an integrated biological system and membrane filter barrier. It is based on the hollow membrane system where microorganisms are attached to the membrane surface, and hydrogen is used as electron donor source. Again, the biomass waste is produced similar to media-based biological processes. In addition, there may be membrane cleaning waste that needs to be treated (pH adjustment) prior to discharge as outlined in Figure 9.9.

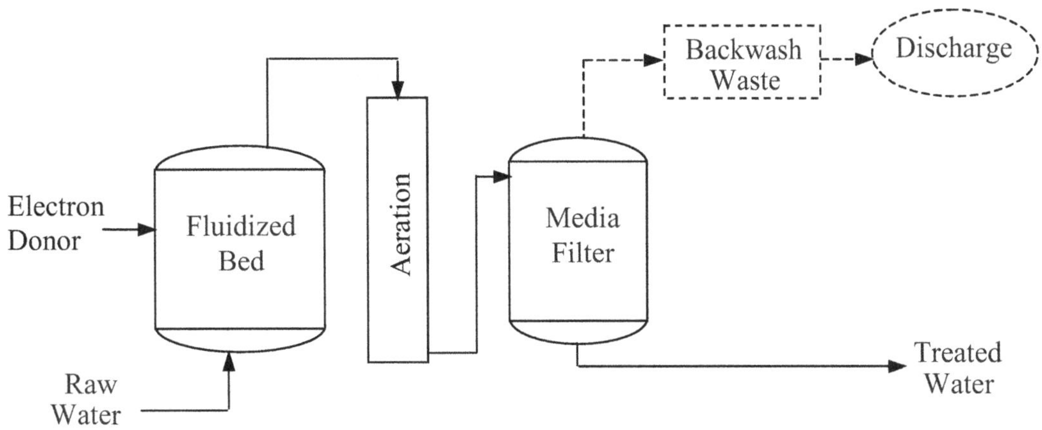

Figure 9.7 Schematic of biological fluidized bed for perchlorate and nitrate treatment

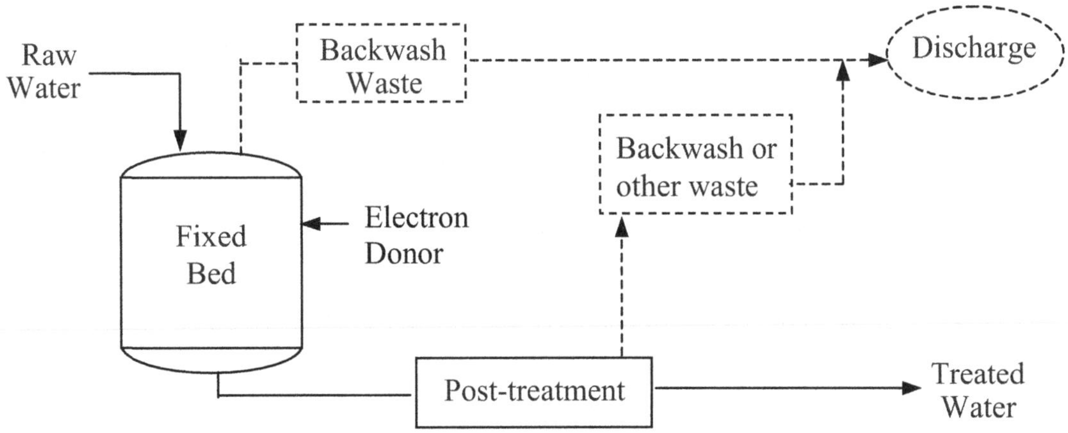

Figure 9.8 Schematic of biological fixed bed for perchlorate and nitrate treatment

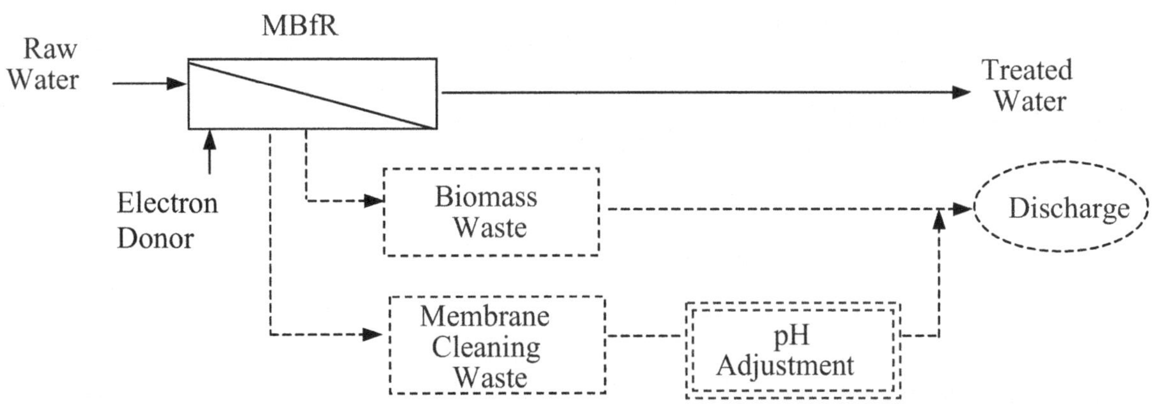

Figure 9.9 Schematic of membrane biofilm for perchlorate and nitrate treatment

Single-use selective ion exchange systems can be applicable for treating perchlorate only. Since nitrate levels in contaminated water are typically 3 or 4 orders of magnitude higher than those for perchlorate, nitrate cannot be treated with this approach as a significant amount of resin will be required to retain nitrate in the resin bed for a long period of time. For perchlorate applications, spent resin is replaced every few months depending on the perchlorate concentration and other anion levels in the raw water. Since the ion exchange process is a phase transfer process, perchlorate is not destroyed, but rather accumulated on the resin, which must be disposed of. During resin replacement, the spent resin is collected and trucked to an incineration facility and destroyed by the equipment provider as part of the resin replacement service. Upon incineration of the spent resin, a certificate of destruction is provided to a utility. There is also a small amount of resin slurry water and start-up waste with high chloride, but these do not contain perchlorate and may be discharged to a sewer depending on local regulations as shown in Figure 9.10.

The conventional, regenerable resin can be used to treat both nitrate and perchlorate. The resin is regenerated with salt brine upon breakthrough of either limiting contaminant (Figure 9.11). Regeneration frequency is on the order of half a day to two days, depending on the operation, and spent brine is discharged to a brine line or collected and trucked off-site for disposal. The amount of spent brine is typically less than 5% of the feed stream. Both perchlorate and nitrate from the resin are transferred to the brine solution during the regeneration process. Depending on the site location and local regulation, this nitrate- and perchlorate-laden spent brine may need to be treated with one of the processes investigated in this project such as biological brine treatment (Chapter 6) or thermal process (Chapter 7). Other waste streams such as backwash or rinse water or spent resin are free from nitrate and perchlorate and can be discharged or disposed of without further treatment.

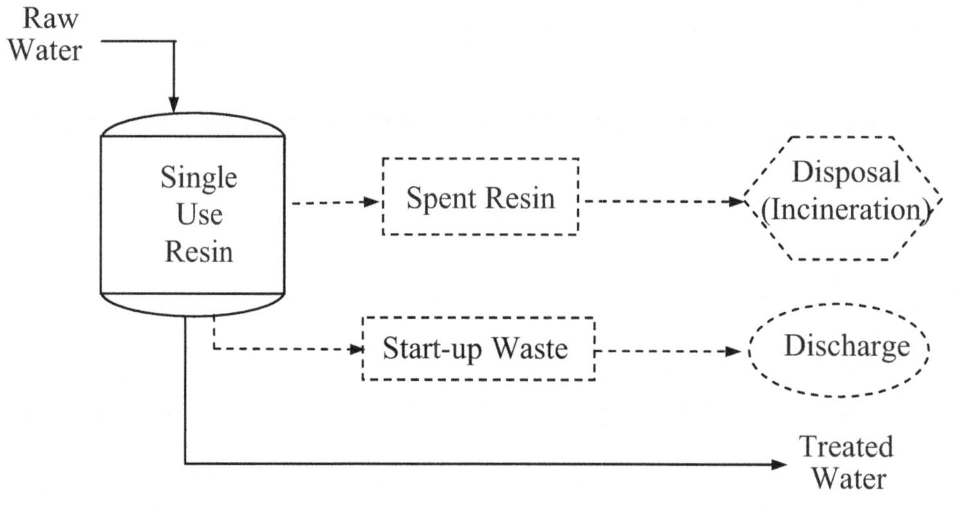

Figure 9.10 Schematic of single-use ion exchange for perchlorate treatment

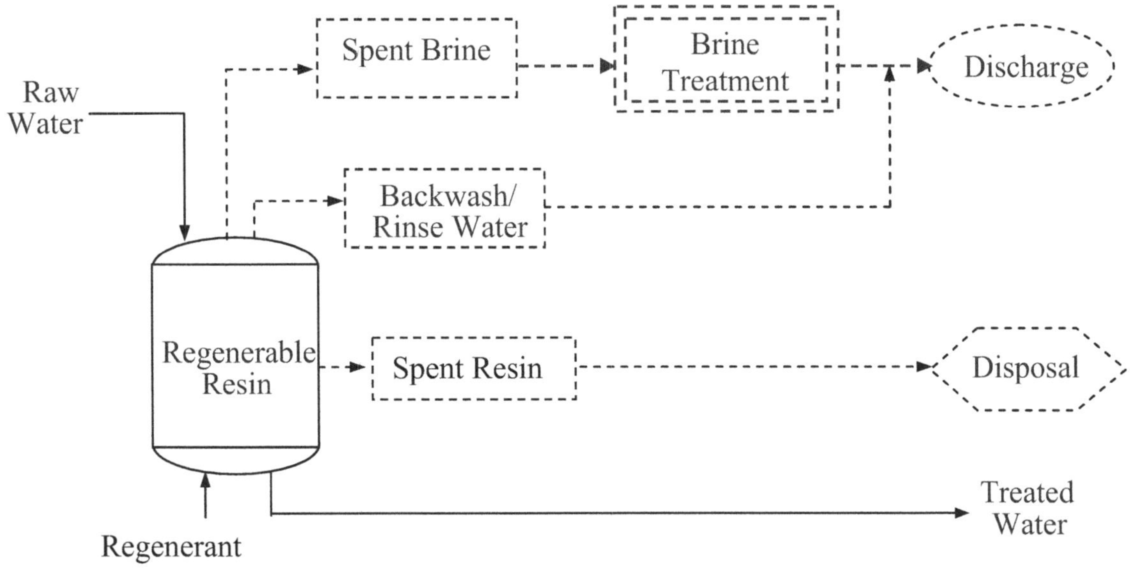

Figure 9.11 Schematic of regenerable ion exchange for perchlorate and nitrate treatment

There has been a new development for using GAC for perchlorate treatment, especially with modified GAC. Virgin or reactivated GAC exhibits limited sorption capacity for perchlorate (a few weeks depending on the perchlorate and nitrate levels), and may be used as a temporary solution. The monomer-modified GAC, on the other hand, is reported to have much higher capacity for perchlorate. Similar to the single-use ion exchange process, GAC or modified GAC retains perchlorate in the GAC bed, and the spent GAC must be disposed of or reactivated off-site on a regular basis (Figure 9.12). This process is still under development and no further information is available on the quantity or characterization of residuals generated from this process at this time.

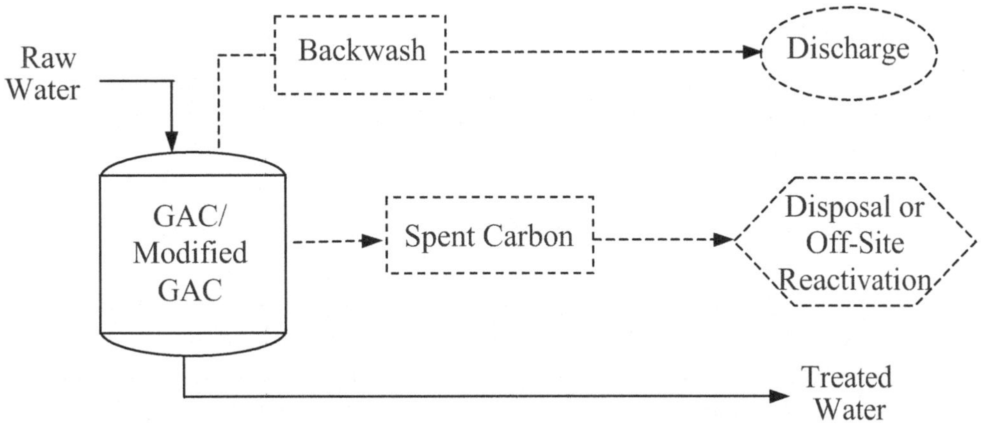

Figure 9.12 Schematic of GAC and modified GAC for perchlorate treatment

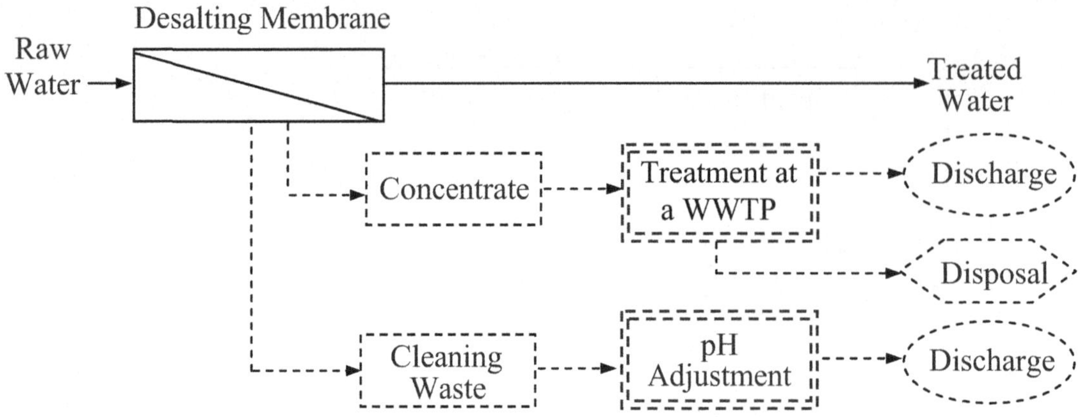

Figure 9.13 Schematic of desalting membrane for perchlorate and nitrate treatment

Generally, membrane processes produce a large quantity (approximately 10 to 25% of feed water as waste) of continuous waste stream containing nitrate and perchlorate. If treatment of waste is required for perchlorate, the cost associated with such application will be high due to the large volume of waste generated. One option to deal with such a large waste is to discharge perchlorate-laden membrane concentrate stream to a wastewater treatment plant and handle perchlorate at a central location, as presented in Chapter 8 of this report (Figure 9.13). Occasional membrane cleaning solution can be discharged to sewer after pH is adjusted.

TREATMENT ALTERNATIVES TO MINIMIZE RESIDUALS

In Chapter 2, a summary of technologies to treat arsenic, perchlorate, and nitrate was provided. Generally, these technologies can be grouped into two categories:

- Water treatment technologies that do not generate waste on-site (e.g., single-use resins or media)
- Residuals treatment technologies to manage waste generated on-site during other treatment processes (e.g. conventional ion-exchange, membrane filtration, or conventional coagulation filtration).

Although the focus of this project was on the second category for treating residuals generated on-site during physical/chemical treatment processes, the following sections provides summary of both categories.

Arsenic - Treatment Technologies Without On-Site Generation of Residuals

For the case of arsenic, the first group of technologies includes single-use media and centrally regenerated media-based treatment systems. These processes do not generate any arsenic-laden liquid waste on-site since arsenic is retained by the sorption media. The arsenic-laden spent media are transported off-site and either landfilled (for single-use media - GFH, GFO, AA, etc.), or regenerated at a central location (centrally-regenerated media - iron-incorporated resin media, TiO_2 media). The new media or regenerated media are then delivered to the site for replacement.

There are a number of such sorbents available commercially or currently under field demonstration phase. Implementation of such technology is expected to have minimum on-site residual issue. This type of technology involves minimum operational requirements and is recommended when on-site discharge is an issue. General treatment approaches using this group of technologies have been studied previously (Chowdhury et al. 2002). Thus, the focus of this study has been on investigating ways to further reduce the small amounts of residuals from backwash water by using prefilter and backwash recycle as presented in Chapter 3. Table 9.1 summarizes this type of technology including maturity of technology, residual reduction of such treatment approach, and potential cost savings.

Table 9.1
Arsenic treatment alternatives with minimum on-site arsenic waste

Technology	Technology maturity	Residual reduction	Cost savings	Technology evaluated in this project
Single-use iron-based granular media (GFH, GFO, etc.)	Commercially available	- On-site liquid waste volume is minimized with these disposable media (compared with other conventional processes) - Spent media is hauled off site and disposed of at the time of media change out	- Simple operation (minimum staffing) - On-site discharge of liquid waste is minimized (occasional waste from backwash once every few weeks to few months)	No
Single-use activated alumina/ modified AA	Commercially available	- Same as above	- Same as above	No
Off-site regenerated media (titanium or zirconium based media, iron-incorporated resin media, etc.)	Commercially available/ Demonstration phase	- Same as above	- Same as above - Cost of centrally regenerated media can be lower than the single-use media	No
Backwash minimization and recycle	Demonstration phase	- On-site liquid waste volume can be further minimized by reducing backwash frequency using pre-filter - Backwash water can be recycled to reduce discharge of liquid waste	- On-site discharge of liquid waste is further minimized	Yes (Ch.3)

Arsenic - Residual Treatment Technologies for Stabilizing Sludge

For large-scale applications, chemical precipitation-based technology is often used. This type of treatment generates arsenic-laden sludge, which is subject to a leaching test to determine the classification for disposal. If the sludge does not pass the leaching test, the sludge is classified as hazardous waste. The waste sludge can be stabilized with lime or cement to reduce leaching of arsenic from the sludge as outlined in Table 9.2. Additional fixatives (e.g., superplasticizer) can also be added as discussed in Chapter 4. The benefit of such a stabilization approach would be a reduced disposal cost due to non-hazardous classification of the spent media as outlined in Table 9.2. The approach for waste minimization is not based on volume reduction but rather on the improvement in quality of the waste to avoid hazardous waste classification as summarized below.

Arsenic - Residual Treatment Technologies for Minimizing Liquid Brine

For phase separation processes, such as ion exchange and desalting membranes (RO, NF, and EDR), concentrated arsenic-laden liquid waste is generated. Chemical coagulants can be added to the brine/concentrate stream to precipitate out arsenic from the liquid waste. This process, however, generates arsenic sludge and may need further treatment as indicated in Table 9.3. An innovative approach investigated in this project is to use hydrogel-based, single-use sorbent to treat waste liquid. The spent sorbent can then be dehydrated and solidified to significantly reduce the waste volume.

Table 9.2
Arsenic sludge treatment alternatives

Technology	Technology maturity	Residual reduction	Cost savings	Technology evaluated in this project
Lime stabilization	Commercially available	- Classification of hazardous waste to non-hazardous waste (improves quality of the waste) - Volume of waste increases due to addition of stabilization chemical	- Cost savings result in disposal cost (from hazardous to non-hazardous) - Cost savings apply to disposal tax, tipping cost, and transportation cost (varies) for non-hazardous waste	Yes (Ch.4)
Cement stabilization	Commercially available/ demonstration phase	- Same as above	- Same as above	Yes (Ch.4)

Table 9.3
Arsenic treatment alternatives to manage arsenic liquid waste

Technology	Technology maturity	Residual reduction	Cost savings	Technology evaluated in this project
Chemical precipitation	Commercially available	- Settles arsenic in liquid brine or concentrate to sludge (waste volume reduction and quality improvement)	- Cost savings result from reduced disposal cost from hazardous to non- hazardous - Cost savings could apply to tipping cost and transportation cost	Yes (Ch.3)
Hydrogel sorbent	Development/ demonstration phase	- Convert liquid brine or concentrate to solid waste (~20 fold reduction in volume and waste)	- Cost savings from volume/weight reduction - Additional cost saving from stabilized arsenic in dehydrated sorbent	Yes (Ch.5)

Perchlorate and Nitrate Treatment Technologies Without On-Site Generation of Residuals

As mentioned in previous sections, a main difference in perchlorate/nitrate residual minimization approach is the fact that these contaminants can be destroyed or mineralized unlike arsenic. Thus the first group of technologies is based on a destructive approach such as biological or high-energy chemical systems to convert perchlorate and nitrate to chloride and nitrogen respectively. A series of AwwaRF and other organization-sponsored projects have investigated such approaches in the past, so these direct treatment technologies were not studied in this project. Table 9.4 summarizes the potential residual reduction and cost savings associated with these technologies that do not generate perchlorate or nitrate waste streams due to the destructive nature of these alternatives.

Another subgroup of technologies that do not produce liquid waste streams on-site is the single-use ion exchange resin or modified GAC-based alternative (Table 9.5) similar to the single-use media discussed for arsenic application. From the residual generation standpoint, these treatment alternatives are equally attractive as treatment alternatives compared to the destruction-based technologies where on-site generation of perchlorate or nitrate waste is minimized.

Table 9.4
Perchlorate and nitrate treatment alternatives - destruction

Technology	Technology maturity	Residual reduction	Cost savings	Technology evaluated in this project
Biological fluidized bed reactor	Commercially available	- Perchlorate and nitrate waste generation is eliminated with this destructive technology compared to other physical separation processes (e.g., ion exchange)	- Disposal or discharge of perchlorate and nitrate residual is eliminated compared with other physical treatment system such as regenerable ion exchange	No
Biological fixed bed system	Regulatory acceptance pending	- Same as above	- Same as above	No
Hollow fiber membrane biofilm reactor	Demonstration phase	- Same as above	- Same as above	No
UV catalysis reduction	Demonstration phase	- Same as above	- Same as above	No

Table 9.5
Perchlorate and nitrate treatment alternatives with no or minimum liquid waste

Technology	Technology maturity	Residual reduction	Cost savings	Technology evaluated in this project
Single-use ion exchange	Commercially available	- On-site waste volume is minimized using this type of disposable media (backwash is not done)	- Unlike regenerable resins, single-use resins do not require a brine line for waste brine disposal	No
Modified GAC	Demonstration phase	- Same as above but backwash is required	- Same as above	No

Perchlorate and Nitrate Residual Treatment Technologies for Minimizing Liquid Waste

The primary focus of this project for perchlorate and nitrate application was on the treatment of the residual waste stream. As indicated in Table 9.6, there are a number of approaches that are applicable to deal with perchlorate/nitrate residual waste streams.

Table 9.6
Residual perchlorate/nitrate treatment alternatives

Technology	Technology maturity	Reduction in residuals	Cost savings	Technology evaluated in this project
Catalytic perchlorate and nitrate destruction module (PNDM)	Commercially available	- Perchlorate and nitrate are destroyed in the ion exchange brine - Volume or salinity of waste brine do not change	- Immediate cost savings is not well defined (pending regulations) - Treated brine may be reused during ion exchange regeneration cycle to reduce salt cost (O&M cost)	No
Thermal/ hydrothermal decomposition	Development phase	- Same as above	- Same as above	Yes (Ch.7)
Biological brine treatment	Demonstratio n phase	- Same as above - If dilution is required for biological treatment at lower TDS, volume of waste can be increased	- Same as above	Yes (Ch.6)
Electro-catalytic conversion of nitrate to nitrogen gas*	Development phase	- Nitrate is destroyed in the ion exchange brine - Volume or salinity of waste ion exchange brine do not change	- Treated brine may be reused during ion exchange regeneration cycle to reduce salt cost (O&M cost)	No
Biological treatment at a WWTP	Development phase	- Perchlorate and nitrate from multiple treatment systems are centrally destroyed - Volume or salinity of waste stream do not change	- Potential cost savings from economy of scale using a central facility to manage local wastes (compared with multiple on-site waste treatments)	Yes (Ch.8)

* Nitrate treatment only

RESIDUAL MANAGEMENT COST CONSIDERATIONS

This section summarizes the cost estimates for arsenic, perchlorate, and nitrate residual treatment technologies, which were investigated in this project. The purpose of this cost estimate is to provide the preliminary budget level numbers to be considered in developing the overall treatment cost. It should be noted that each cost estimate differs in the way that is projected for treating different types and quantities of waste. For example, the cost estimate for biological treatment of perchlorate and nitrate brine is based on conventional ion exchange where sodium chloride is used, whereas the estimate for thermal treatment of brine is for a different type of regeneration where ferric chloride and hydrochloric acid are used to significantly reduce the volume of brine waste. Similarly, the cost estimates presented for arsenic residual management also deal with different types of waste streams (e.g., backwash water, sludge, and ion exchange brine). Thus, the cost estimates presented in this section are specific to an integrated treatment approach based on different units (e.g., sludge vs. liquid) and are not designed to be compared to each other.

Arsenic Backwash Minimization Cost

Pretreatment for solids removal has the potential benefit of reducing the need to backwash the arsenic-adsorption media and therefore reduces the amount of spent backwash requiring disposal. The capital and O&M cost estimates for a 100-gpm groundwater plant can be determined by assuming a filtration flow of about 1 gpm/EQSL (a rate about 4 times lower than manufacturer-recommended, based on pilot test results), four 8-in diameter filter housings (about $4,000 each), each containing six 10-mm cartridge filters of 4 EQSL (about $30 each). The initial capital costs, including installation costs, would be about $20,000. With an assumed cartridge filter replacement on a 3-month frequency the annual O&M costs would total about $3,000.

Arsenic Backwash Treatment Cost

Based on the pilot testing results, filtration of spent backwash by a cartridge filter would not be able to handle backwash operation upsets (e.g., too much media particle carry-over). Coagulation-sedimentation would be a more reliable option, although it would require a skilled operator (O&M costs may increase if such staff is not readily available at the utility). Capital and O&M costs for coagulation-sedimentation are shown below, based on the following assumptions:

- 100-gpm groundwater system.
- Standard 3-vessel system.
- 5-min empty bed contact time (EBCT).
- loading rate of 5 gpm/ft^2 for GFH media.
- Backwash rate of 10 gpm/ft^2 for 15 minutes.
- Vessels backwashed during the same event.
- Monthly backwash frequency.
- 20 mg/L ferric chloride.
- 20-min flocculation time.

The media surface loading area is then 20 ft^2, which means that 3,000 gallons of spent backwash water is generated at each backwashing event. The spent backwash can be treated by coagulation-sedimentation in a batch approach using a 400-ft^3 tank (8-foot high, 8-foot diameter). Ferric chloride feeding and storage equipment, a mixer (for rapid mix and coagulation), and pumps for the supernatant (return flow) and the residuals. Sedimentation residuals may need to be stabilized or disposed of in a hazardous landfill, which may increase the O&M costs (see next section for cost details about residuals stabilization).

Capital costs (installation costs included) are estimated to be about $10,000 (this may vary depending on tank material type). Annual O&M costs for ferric chloride will be negligible since only 15 lbs of a 40-percent ferric chloride solution would be required to treat 3,000-gal spent backwash water 12 times a year.

Arsenic Residual Stabilization Cost

Based on preliminary tests shown in Chapter 4, arsenic-containing ferric residuals (40% solids) could be stabilized with about 15% (w/w) lime. Assuming arsenic-containing ferric residuals can be dewatered to this level (either from lagoons and drying piles like LADWP or by mechanical equipment) then a dry lime feed system (small utilities could just use 25-lb bags and a small scale), a lime-sludge mixer, and a small front-end loader to move the sludge cakes.

Non-hazardous disposal cost is estimated at about $60/ton, with a $40/ton tipping fee on average and a hauling fee of about $20/ton. Actual hauling fees will vary depending on the distance to a designated disposal facility. The cost of hydrated lime can vary significantly among various U.S. regions, but is on average about $90/ton. Lime-conditioned residuals would be about 15% heavier than untreated residuals, thereby adding to the cost of landfill disposal (tipping and hauling).

Hazardous landfill disposal would be for untreated, arsenic-containing ferric residuals that fail the CaWET test. The estimated cost is about $300/ton and includes tipping fee, hauling fee, separation and handling costs, and various special local and state taxes. A small front-end loader would be necessary to transport the residuals cakes (40% solids) to the hauling trucks.

To make lime stabilization cost competitive with hazardous landfill disposal of untreated residuals, the O&M cost of lime (the power cost is assumed to be negligible compared to lime cost) and the annualized costs of the dry lime feed system and lime-sludge mixer must remain below $300/ton.

Stabilized sludge will be heavier by 15% (from the lime addition) so the non-hazardous landfill disposal cost becomes ($60/ton x 1.15) = $69/ton. The cost of lime per ton of residuals is (15% x $90/ton[lime]) = $13.5/ton[residuals]. Hence the annualized costs for the equipment would have to be $300/ton - $69/ton - $13.5/ton = $217.5/ton. Assuming 8% interest and 20-year term for the annualized costs, then the capital cost for the equipment need to remain below $2,135/ton for the stabilization option to remain cost competitive with hazardous landfill disposal.

Arsenic Residual Solidification Cost

The hydrogel sorbent tested for arsenic solidification is not commercially available. Thus, it is difficult to project the cost of such media. However, it is expected that the cost will be comparable to other types of sorbent since the technology is based on readily available chemicals and the synthesis process requires a similar level of production as manufacturing of ion exchange resins. Due to its unique characteristics, dehydrated spent sorbent will reduce its size by 20 times (or 95% reduction in volume and waste) from its original hydrated form. It is anticipated that the combination of high arsenic to iron sorption ratio and waste volume reduction after dehydration will justify the media cost for this option to be a viable approach.

Perchlorate and Nitrate Brine Biodegradation Cost

Due to the site-specific conditions, design basis must be developed for each specific application. Capital costs are most affected by system residence time and the need for dilution. Shorter residence time results in a smaller footprint and smaller tanks and equipment. Dilution of brine increases treatment volume, which affects capital cost and operating cost for dilution water, nutrient, electrical power, and effluent disposal. Table 9.7 provides a summary of conservative design basis for a suspended growth biodegradations process without biomass retention. Other studies have shown that much higher salt concentrations (6%) can be effectively treated at relatively short residence times (<12 hours) by retaining biomass.

Treatment costs were estimated according to the design basis in Table 9.6 and the following assumptions:

- Brine volume generated equals ~1% of groundwater volume treated.
- Undiluted brine treatment rate is 25 gpm based on La Puente, Calif. scenario.
- Brine is diluted to 30% concentration – 1,000 gallons of brine yields a 3,333 gallon diluted process stream.
- Nutrient cost $0.73/gallon delivered and includes micronutrients.
- Caustic cost $1.78/gallon delivered.
- Dilution water cost $0.50/1,000 gallons.
- Electricity includes heating costs at $0.08/kWhr.
- Treated brine disposal costs include surcharges for volume, COD, and TSS based on Los Angeles County Sanitation District rates: $0.50/1,000 gallons volume, $0.09/lb of COD, $0.25/lb of TSS could add another $4.50 per 1,000 gallons of undiluted brine.

Table 9.7
Typical design basis of biodegradation process without biomass retention

	TDS feed (percent)	HRT (hr)	Temperature (°C)	pH	Nutrient (vol percent)	Caustic (vol percent)
Minimum	1.0	8	20	6.5	0.3	0.1
Maximum	3.0	24	40	8.0	0.5	0.3
Design	2.0	16	30	7.0	0.4	0.2

Table 9.8
Summary of brine biodegradation costs

Parameter	Per 1,000 gallons of brine	Per acre-foot of groundwater
Nutrient	$2.92	$9.52
Caustic	$3.56	$11.61
Dilution Water	$1.17	$3.80
Electricity	$0.44	$1.45
Disposal		
Volume	$1.65	$5.83
COD	$2.25	$7.34
TSS	$0.75	$2.45
Total Cost:	$12.74	$41.54

A summary of the treatment costs is provided in Table 9.8. This cost estimate is very conservative and design improvements currently being evaluated have the potential to significantly reduce these treatment costs. For a La Puente-sized treatment system labor could add approximately $20 per acre-foot and depreciation (30-year plant life) could add another $20 per acre-foot.

Perchlorate and Nitrate Thermal Destruction of Spent Ferric Chloride Regenerating Solution

Thermal treatment cost is directly related to the volume of spent regenerating solution treated, which is related to regeneration efficiency and resin treatment capacity. The actual treatment capacity of ion exchange resin is highly dependent on water quality and treatment level. Capacity can range from 20,000 to 40,000 bed volumes for high TDS (>1,000 mg/L) groundwater, to 200,0000 to 400,000 bed volumes or greater for low TDS (<50 mg/L) groundwater.

The ferric chloride regeneration analysis is based on the following assumptions:

- Resin treatment capacity of 50,000 bed volumes for perchlorate-specific resins.
- Net ferric chloride regenerating solution consumption of 2 bed volumes - Based on reusing the regenerating solution several times, and spent regenerant volume is 0.004% of the treated ground water.
- Ferric chloride solution cost $0.75/gallon - NSF certified ferric chloride ($400/dry ton) and NSF certified concentrated HCl ($105/ton, $0.12/Kg).

Table 9.9
Ferric chloride regenerating solution treatment and replacement cost

Component	Quantity	Unit Cost	Cost per Acre-Foot
Regenerating solution	13 gal	$0.75/gal	$9.75
Electricity	4 kWhr	$0.10/kWhr	$0.40
Natural gas	0.16 M Btu	$12/MBtu	$1.92
Iron oxide credit	4 Kg	$0.20/Kg	($0.80)
HCl credit (18% HCl)	80 Kg	$0.04/Kg	($3.20)
Total Cost:			$8.07

An analysis of the cost of ferric chloride regeneration and destruction of spent solution generated from the treatment of one acre-foot of groundwater is provided in Table 9.9. Costs are based on spray roasting data and assumption provided above.

Comparison to ISEP Technology with Sodium Chloride Regeneration

As a comparison, a cost estimate for commercially available ISEP system for perchlorate and nitrate brine is provided, with the following assumptions.

- Brine regenerating solution consumption typically is 1% to treated groundwater.
- Sodium chloride brine cost $0.016/gallon - Brine regenerating solution is 7% sodium chloride (NaCl) and salt cost is $55/ton.

Therefore, the salt cost to treat one acre-foot of water is over $50.

(326,000 gal/acre-foot) x (1%) x ($0.016/gal) = $52.16/acre-foot.

This cost does not include treatment to destroy the nitrate and perchlorate in the brine or any reuse or recovery credit and associated treatment cost. High temperature catalytic treatment or biological treatment of the spent NaCl brine can only recover 80 to 90% of the chloride because a portion of it is consumed during ion exchange with anions in the groundwater and some will be lost in slip streams that are necessary to prevent excessive levels of sulfate and other materials. Therefore, the minimum salt makeup for NaCl brine regeneration with salt recovery is $5 to $10 per acre-foot of water treated. This cost does not include the cost of treatment or recovery.

Perchlorate and Nitrate Treatment Cost at a WWTP

Capital costs for treating perchlorate or nitrate residuals at a WWTP would involve the installation of fixed-bed reactors and associated backwash pipework, valves, and air scour system. A concrete pad and GAC would also be required. No provisions need to be made to treat backwash wastewater, as it could be discharged directly to the secondary treatment processes.

Capital costs (including installation) are estimated at approximately $1.5 million for a 3-mgd fixed-bed treatment unit installed ahead of secondary treatment at a WWTP ($0.50/gpd). The following additional assumptions apply: EBCT = 15 minutes, feed perchlorate concentration less than 2 mg/L, feed DO concentration less than 0.5 mg/L, and feed nitrate concentration less than 3 mg/L.

Annual O&M costs are estimated at approximately $18,000 for a 3-mgd fixed-bed treatment unit installed ahead of secondary treatment at a WWTP (less than $0.02/1,000 gallons). O&M costs include power costs and it is assumed that backwash waste is discharged to the secondary treatment process. It is also assumed that background organic constituents in the wastewater serve as the electron donor for biological reduction processes. This cost estimate is based on the configuration (fixed bed) used in this report, and the cost estimate will be different when alternative support media is used to sustain biological growth.

OPERATIONAL AND REGULATORY CONSIDERATIONS

Type, quantity of waste generation, disposal method, and permit requirement are site-specific, and it is difficult to project such information without detailed site/location constraints. To help utilities with such information, a number of case studies with full or demonstration scale plants treating arsenic, perchlorate, or nitrate are provided in this section. A summary of case studies for arsenic plants and perchlorate/nitrate plants are provided in Table 9.10 and Table 9.11 respectively. For each utility, the plant capacity, the concentrations of contaminants, the type and quantity of wastes, and the disposal methods are outlined. In addition to information on residual production, any special permit requirements are also included in the summary table. The goal of this utility survey was to provide the full scale plant information that will provide more insight into residual management issues.

DECISION TREES FOR SELECTING TREATMENT TECHNOLOGIES WITH RESIDUAL MANAGEMENT

Selection of the most appropriate technology is a challenge where a number of factors need to be considered. In general, the capital and O&M costs are the most important criteria in making a treatment process selection. However, there could be other critical decision criteria that need to be incorporated, such as project-specific site limitations, staffing requirements, or other unique circumstances. One of these critical decision criteria is residual treatment and management cost, which is sometimes not incorporated in the treatment process selection. The following is a list of criteria often considered for treatment process selection:

- Capital cost.
- O&M cost.
- Residual production.
- Waste disposal and discharge options.
- Water quality.
- Future Regulations.
- Flow rate.
- Central vs. non-central.
- Project location.
- Chemical requirement.
- Automation.
- Operator level.
- On-line analysis.
- Footprint.

Table 9.10
Case studies for utilities with full-scale arsenic treatment facilities

Utility	Treatment type	Capacity (gpm)	Arsenic concentration	Waste Type	Quantity	Waste handling	Permits
A (NC)	Membrane (RO)	2,100	48-58 µg/L	Liquid concentrate	~850,000 gal/day	Concentrate discharge to ocean	NPDES
B (AZ)	Single-use Media (GFO)	1,700	32 µg/L	Backwash and spent media	N/A	Sewer Discharge	None
C (CA)	Conventional Treatment	312,500	25 µg/L	Sludge	18,000 tons/year	Landfill	Disposal
D (ME)	Regenerable ion exchange	4.5	~13 µg/L	Spent brine	NA	Septic	None
E (AZ)	Single-use Media (GFH)	1,100	~17 µg/L	Backwash and spent media	~81,000 gal/day	Sewer Discharge	None
F (CA)	Regenerable ion exchange	2,100	13~15 µg/L	Spent brine	~150,000 gal/day	Hold and haul to a local POTW	NPDES

NA: Information Not Available

Table 9.11
Case Studies for utilities with full-scale perchlorate or nitrate treatment facilities

Utility	Treatment type	Capacity (gpm)	Contaminants (concentration)	Waste type	Quantity	Waste handling	Permits
A (CA)	Regenerable ion exchange	9,000	Nitrate (31 – 84 mg/L) Perchlorate (~14 μg/L)	Liquid Brine	~250,000 gal/day of brine	Brine discharge	NA
B (CA)	Regenerable ion exchange	2,500	Nitrate (15 mg/L) Perchlorate (35 μg/L)	Liquid Brine	25 gpm of brine	Brine discharge to until 2006 Perchlorate brine treatment by 2006	Perchlorate Discharge
C (CA)	Single-use ion exchange	15,000	Perchlorate (60μg/L)	Spent resins and wash	Resin replacement every 6 weeks	Spent resin incineration and wash water discharge	Discharge NPDES
D (CA)	Regenerable ion exchange	1,350	Nitrate (45 mg/L)	Liquid brine	1,700 gal/day of brine	Brine trucking to a WWTP by the equipment provider	NPDES pending Brine permit to WWTP
E (CA)	Single-use/ regenerable ion exchange	2,000	Nitrate (50 mg/L) Perchlorate (15 μg/L)	Spent resin, wash water, and liquid brine	Resin replacement every 5 month, 1,500 gal/day of spent brine	Resin incineration and brine trucking to a WWTP by the equipment provider	NPDES pending Brine permit to WWTP

(continued)

Table 9.11 (Continued)

Utility	Treatment type	Capacity (gpm)	Contaminants (concentration)	Waste type	Quantity	Waste handling	Permits
F (CA)	Single-use ion exchange	4,000	Perchlorate (35 μg/L)	Spent resin and wash water	Resin replacement every few months	Resin incineration and Wash water discharge to sewer	NA
G (CA)	Regenerable ion exchange	1,000	Nitrate (50 mg/L)	Liquid Brine	6,000 gal/day	Brine discharge to a WWTP (pending)	NPDES Discharge (pending)
H (CA)	Regenerable ion exchange	400-2,100	Nitrate (55 – 70 mg/L)	Liquid brine	140,000 gal/day	Brine discharge to a WWTP	NA
I (CA)	Regenerable ion exchange	3,000	Nitrate (34 - 80 mg/L)	Liquid brine	6,000 gal/day	Regional WWTP	Discharge-Batch
J (CA)	Regenerable ion exchange	7,900	Perchlorate (NA)	Liquid brine	80 gpm of perchlorate brine	Brine discharge until 2006 Brine treatment by 2006	NA
K (CA)	Membrane/ Regenerable ion exchange	6,250	Nitrate (50-150 mg/L)	Concentrate and Brine	NA	Concentrate and brine discharge	NA
L (CA)	Biological Fluidized Bed	6,000	Perchlorate (4-6 mg/L)	Waste Biomass	NA	Sewer Discharge	NA

NA: Information Not Available

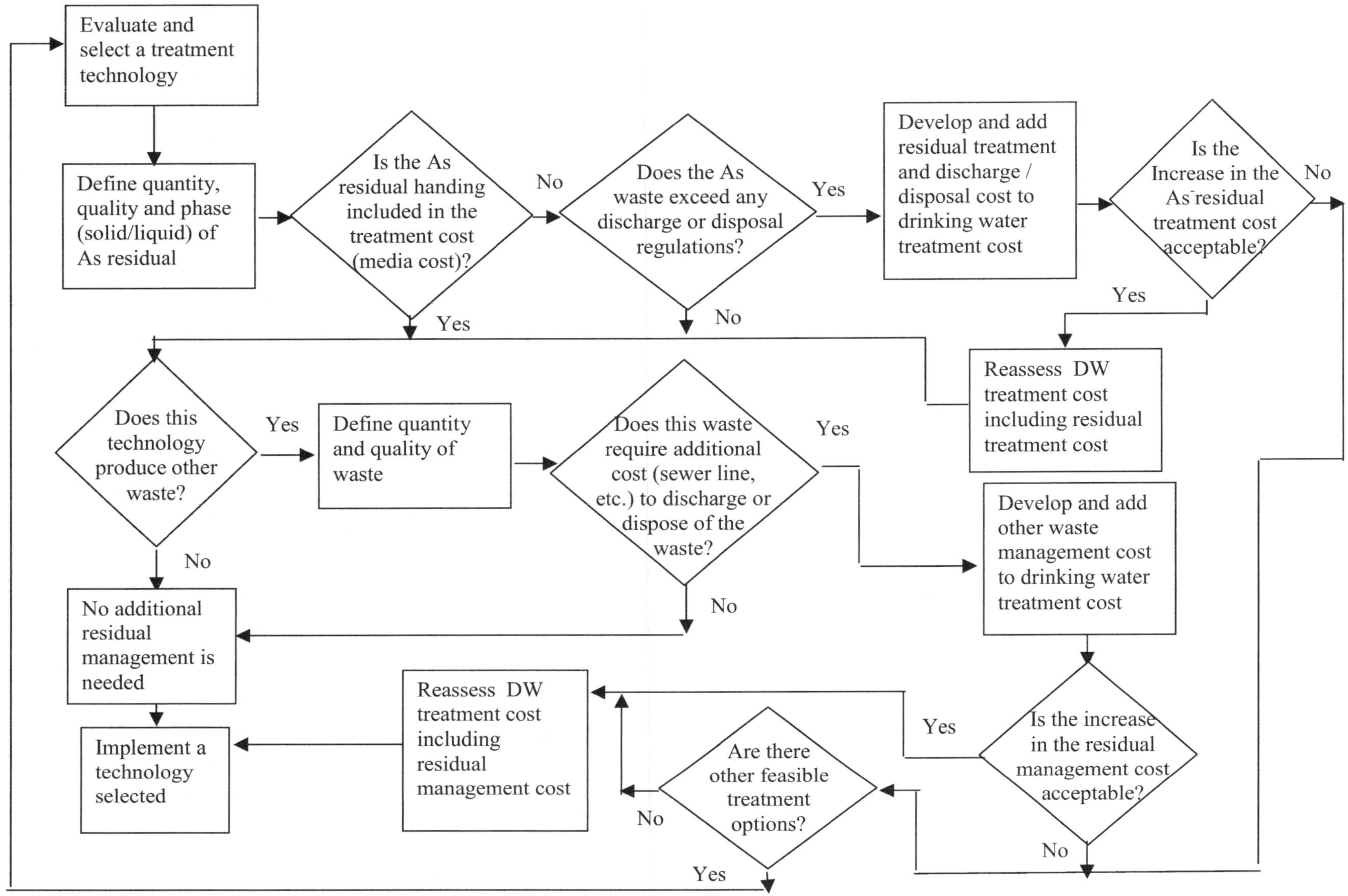

Figure 9.14 Arsenic technology implementation decision tree including residual management

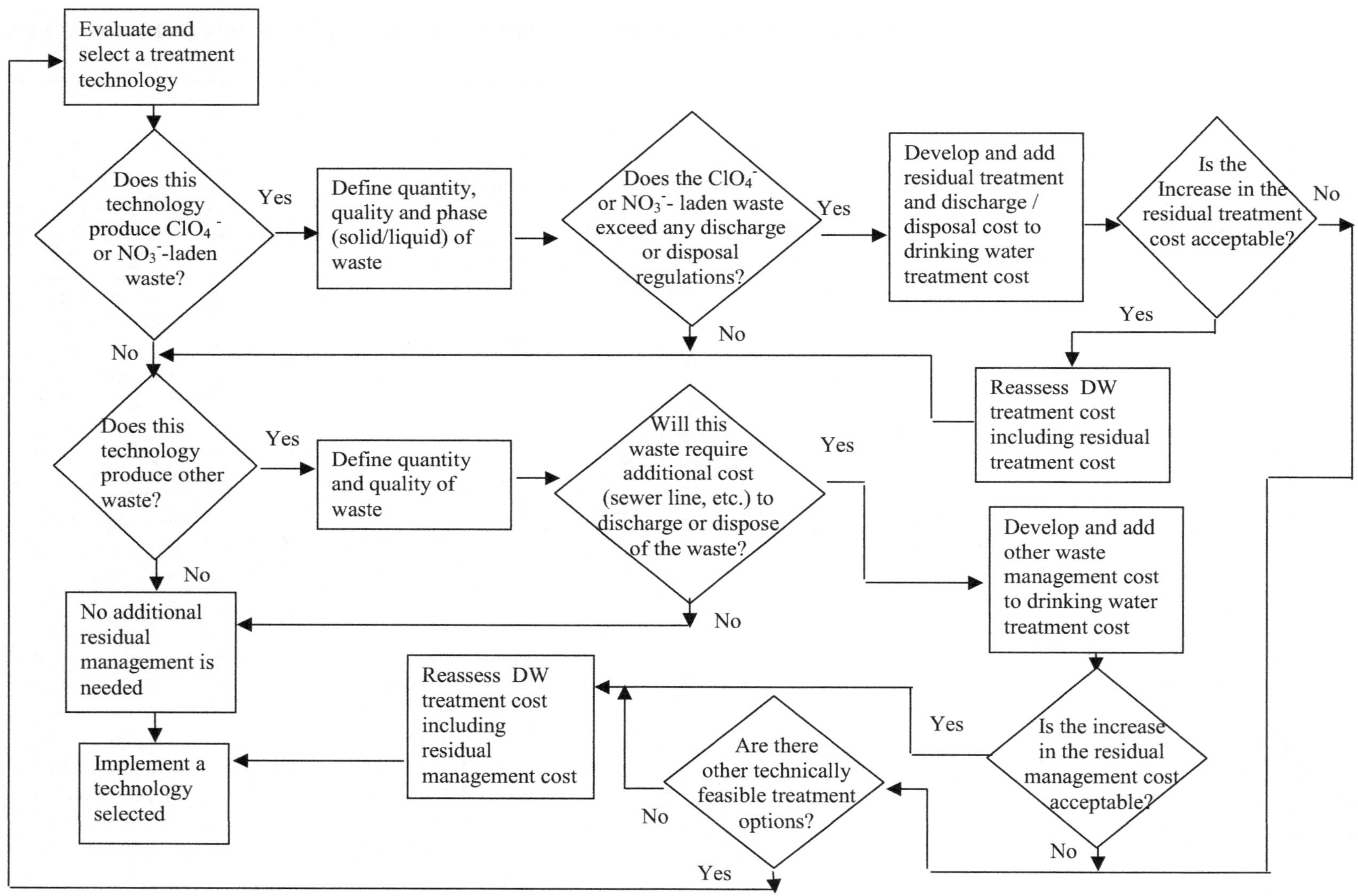

Figure 9.15 Perchlorate and nitrate technology implementation decision tree including residual management

One approach in integrating the key criteria to select the best treatment process for arsenic, perchlorate, and nitrate is to use a decision-making tool to prioritize the issues and define constraints. There are other tools that provide recommendations for arsenic treatment such as the one developed for AwwaRF (Chowdhury et al., 2002). The intent for the following decision trees is to provide a guidance and to point out various factors that need to be considered in making process selection. Figure 9.14 shows an implementation decision tree including the residual management aspect by reiterative process to evaluate whether a selected drinking water treatment process is a justifiable alternative when residual management is included. A similar flow diagram for perchlorate and nitrate is shown in Figure 9.15 as a guidance tool.

SUMMARY AND RECOMMENDATIONS TO UTILITIES

Selection of the most appropriate technology is a challenge where a number of factors must be considered as discussed. A treatment technology that provides good treatment performance in one location may not be suitable for other locations due to the limited discharge options, or distance to disposal facility, etc. This project investigated treatment alternatives for residuals from arsenic, perchlorate, or nitrate treatment processes. Some of the results presented in the report are based on the technologies that are already available while new and innovative methods and concepts have also been demonstrated during the project. An important aspect of this project was to collect and disseminate information on residual treatment, a step which is often overlooked when making a process selection.

In general, water quality for potential constituents needs to be thoroughly evaluated even if their levels may be lower than the drinking water guidelines. Many technologies concentrate contaminants in waste streams, and this may result residual levels that exceed discharge or disposal regulations. As such, a review of the technologies are being considered should not only focus on the drinking water treatment side, but also on the residual management side to avoid any additional cost for residual management. In order to effectively implement a treatment system, understanding the local, state, and federal drinking water regulations (current and future) as well as discharge and disposal guidelines are critical steps. It is also important to recognize that each project is unique and that overall arsenic, perchlorate, and nitrate management requires a multifaceted approach that integrates technology, cost, regulation, and the public.

Specifically based on the results of residuals management technologies evaluated in this project, the following recommendations are made to utilities considering implementation of residual treatment alternatives for arsenic, perchlorate, and nitrate.

- Pre-filtering raw water to minimize backwash frequency from single-use media treatment for arsenic could extend the service run time and minimize backwash frequency to reduce waste discharge. The actual reduction in backwash frequency will depend on the quality and constituents in the raw water. It is recommended, however, that the cost of pre-filtration should be compared with the cost of treating the spent backwash. For treating backwash streams, the cartridge filters may be difficult to implement in backwash water treatment due to rapid headloss buildup caused by media carry over from some granular iron-based systems. As such, further study of the coagulation-sedimentation process for backwash water from single-use sorbent is recommended.
- Arsenic-containing sludge was more effectively stabilized using lime based on the results from this study and previous research (Cornwell et al. 2003). It is recommended that lime be used for stabilizing arsenic-containing residuals from coagulation sedimentation processes rather than cement (with or without

superplasticizer). Lime is more cost competitive than cement and the water industry is already familiar with lime additions of up to 20% (w/w) and using filter presses for dewatering water treatment plant residuals. Furthermore, elevated lime concentrations in the stabilized sludge will be valuable to the farming community if the stabilized residuals are land applied.

- The hydrogel sorbent used for solidifying arsenic brine waste showed sorption capacity as much as 1 g of arsenic per 1 g of iron under the conditions tested. As indicated previously, in coagulation much of the iron is in the bulk material of the iron oxyhydroxide floc and is unavailable for arsenic binding. With the Ca-Fe gel beads, arsenic is adsorbed throughout the entire volume of the bead and a high proportion of the iron is available to bind arsenic. In contrast to coagulant sludge, the spent beads are easily dehydrated, producing a material that is amenable to handling and transport. The spent, dehydrated beads are compatible with the TCLP test but are not stable under conditions of the CaWET test. Thus, in California, some additional stabilization of the spent, dehydrated beads would be required prior to disposal. However, further testing of the beads, at pilot-scale, would be required to assess possible impacts of bead stability on performance over longer periods.
- High TDS conditions reduced biomass production and decreased the reduction rates for both nitrate and perchlorate. Therefore, it is necessary to maintain modest brine concentrations (2 to 3%) and moderately long residence times (12 to 16 hours) in suspended-growth, stirred-tank-reactor systems. Brine biodegradation conducted under this project did not address recovery and recycle of biomass to improve performance. However, other ongoing studies have shown that suspended growth processes, with mechanisms to retain biomass, significantly improved performance and enabled the biodegradation of brine at much higher concentrations. Biomass recovery would also reduce disposal costs that may be levied on treated waste effluent. It is recommended that perchlorate biodegradation of this brine be evaluated using a biodegradation system that recovers and reuses biomass. This can be accomplished by using technologies such as sequencing batch, which was demonstrated in this study, fixed film bioreactors, or adding a clarifier or a membrane filtration step. Biomass concentration and reuse should improve perchlorate biodegradation performance in addition to minimizing effluent disposal and system operation costs.
- Another technology to greatly reduce perchlorate (and nitrate) waste is to use a combination of selective resin and ferric chloride / hydrochloric acid-based regeneration technology. Based on the testing conducted on ferric chloride residuals, two technologies were identified that could be implemented to achieve perchlorate destruction and partial recovery and reuse of components. These commercially available technologies are spray roasting and spray drying. It is recommended that alternative regeneration technologies using ferric chloride / hydrochloric acid for selective resin be considered and further evaluated. Such follow-up work will provide additional information on the potential for decreased volume of generated wastes as well as recovery and reuse of regenerant solution components.
- Perchlorate- and nitrate-laden concentrates from desalting membrane-based plants combined with municipal wastewater may effectively be treated using a fixed-bed biological process as demonstrated in this project. The concept can also be applied as a regional solution where a number of such waste streams can be centrally treated. However, it is recommended that additional research be performed to resolve the

remaining unknowns related to the design, start-up, and operation of such a process. Some of these unknowns include media selection, backwash protocol, impact of intermittent perchlorate exposure on degradation kinetics, and temperature effects. It is also recommended that studies be performed to investigate the efficacy of using a bioreactor to treat concentrate streams that have been blended with scalped municipal wastewater, an approach that might be more cost-effective than treating the entire wastewater flow.

REFERENCES

Abu-Omar, M.M., L.D. McPherson, J. Arias, and V.M. Béreau. 2000. Clean and Efficient Catalytic Reduction of Perchlorate. *Angew. Chem. Int. Ed.* 39(23): 4310-4313.

Attaway, H. and M. Smith. 1994. Propellant Wastewater Treatment Process. U.S. Patent 5,302,285.

Attaway, H. and M. Smith. 1993. Reduction of Perchlorate by Anaerobic Enrichment Culture, *J, Ind. Microbiol.*, 12:408-412.

Bonder, M., H. Mell, E. Stupperich, and A. Kroger. 1982. Biosynthetic Pathways of *Vibrio succinogenes* Growing with Fumarate as a Terminal Electron Acceptor and Sole Carbon Source, *Arch. Microbiol.*, 131:216-223.

Brierley, J.A. 1990. Biotechnology for the Extractive Metals Industries. *JOM. J Min. Met. Mat. S.* 42(1): 28-30.

Brown, J.C., V.L. Snoeyink, and M.J. Kirisits. 2002. Removal of Perchlorate at Low μg/L Concentrations by Sorption and Biological Reduction in an Activated Carbon Filter, *Journal AWWA,* 94(2): 70-79.

Calgon Carbon Corporation 2002. WRS, ISEP Presentation, January 2002.

Cannon, Fred S. and Chongzheng Na. 2000. "Perchlorate Removal Using Tailored Granular Activated Carbon and Chemical Regeneration" in Perchlorate Treatment Technology Workshop, 5th Annual Joint Services Pollution Prevention & Hazardous Waste Management Conference & Exhibition, August 21-24, 2000, Henry B. Gonzalez Convention Center, San Antonio, Texas.

Carbonell-Barrachina, A.A., A. Jugsujinda, F. Burlo, R.D. Delaune, and W.H. Patrick. 2000. "Arsenic chemistry in municipal sewage sludge as affected by redox potential and pH." *Wat. Res.,* 34:1, 216-224.

Catts, J. 1998. Perchlorate Stake Holder's Forum, Henderson, NV.

Chowdhury, Z., S. Kommineni, R. Narasimhan, J. Brereton, G. Amy, S. Sinha. 2002. *Implementation of Arsenic Treatment Systems Part 1. Process Selection.* Denver, Colo.: AwwaRF and AWWA.

Chen, A.S.C., K.A. Fields, T.J. Sorg, and L. Wang. 2002. Field Evaluation of As Removal By Conventional Plants. *JAWWA.* 94:9:64.

Chen, D., Z. Lewondowski, F. Roe, and P. Surapaneni. 1993. Diffusivity of Cu^{2+} in Calcium Alginate Gel Beads. *Biotechnol. Bioeng.* 41(7):755-760.

Chwirka, J. 1999. Residuals Generation, Handling and Disposal. In *Arsenic Treatment Options and Residuals Handling Issues.* Draft Final Report. AWWA, Denver.

Clesceri L. S., A.E. Greenberg, and R.R. Trussell. 1989. Standard Methods for the Examination of Water and Wastewater. 17th Edition: (3) 102-106.

Clifford, D. and C.C. Lin. 1995. Ion Exchange, Activated Alumina, and Membrane Processes for Arsenic Removal from Groundwater, *Proceedings of the 45th Annual Environmental Engineering Conference.* University of Kansas, KS.

Clifford, D. and C.C. Lin. 1986. Arsenic Removal From Groundwater in Hanford, California - A Summary Report. Houston, Texas: University of Houston.

Clifford, Dennis, G. Ghurye, and A. Tripp. 1998. Arsenic Ion Exchange Process with Reuse of Spent Brine. *In Proc. of 1998 Annual AWWA Conference.* Denver, Colo.: AWWA.

Coates, J.D., U. Michaelidou, R.A. Bruce, S.M. O'Connor, J.N. Crespi, and L.A. Achenbach. 1999. Ubiquity and Diversity of Dissimilatory (Per)chlorate-Reducing Bacteria. Applied Environmental Microbiology, December: 5234-5241.

Coppola, E.N. and G.R. McDonald. 2000. Biodegradation of Ammonium Perchlorate, Nitrate, Hydrolysates, and Other Energetic Materials. U.S. Patent 6,077,432.

Coppola, E.N., J. Rine, and G. Startzell. 1998. Operational Implementation of Ammonium Perchlorate Biodegradation, AFRL-ML-TY-TR-1999-4524, June 19, 1998.

Coppola, E.N., J. Rine, and G. Startzell. 2003. Optimization of the ARA Perchlorate Biodegradation System at ATK-Thiokol Propulsion, Tech Report, Air Force Research Laboratory..

Crusberg T.C., P.J. Weathers, and E.F. Baker. 1991. Biotraps for Heavy Metal Removal and Recovery from Industrial Wastewaters, *Biological Processes* Vol.3: Innovative Hazardous Waste Treatment Technology Series, Freeman, H.M., and P.R. Sferra Eds. 91-96.

Deshmukh S. et al. 2002. High Energy Electron Injection Pilot Test. Studies conducted at the Orange County Water District. . CEC Draft Final Report.

Draget, K.I., K. Steinsvacg, E. Onsoyen, O. Smidsrod. 1998. Na and K-Alginate: Effect on Ca^{2+} Gelation. *Carbohyd. Polym.* 35(1-2):1-6.

Driehaus, W., M. Jekel, U. Hildebrandt. 1998. Granular ferric hydroxide – a new adsorbent for the removal of arsenic from natural water. J. Water SRT – Aqua 47:1.

Early Sr., J.E., D.C. Tofan, and G.A. Amadei. 1999. Rapid Reduction of Perchlorate Ion by Titanous Ion in Ethanolic Solutions. Paper presented at the ACS National Meeting, New Orleans, LA, August 22-26, 1999.

Ela, W.P., R. Sierra-Alvarez, J.A. Field, A. Ghosh, I. Cortinas, and C. de las Casas. 2003. Laboratory Tests on the Fate of Arsenic in Landfills. *Proceedings of the 2003 Residuals and Biosolids Symposium*. Baltimore, MD: AWWA and WEF.,

Fields, Keith, A. Chen, and L. Wang. 2000. Arsenic Removal from Drinking Water by Iron Removal Plants. EPA/600/R-00-086. Cincinnati, OH.

Fundueanu, G., C. Nastruzzi, A. Carpov, J. Desbrieres, and M. Rinaudo. 1999. Pysicochemical Characterization of Ca-Alginate Beads Microparticles Produced with Different Methods. *Biomaterials*. 20(15):1427-1435.

Gallup, D.L. 2002. UNOCAL, personal communication

Giblin, T., D. Herman, M.A. Deshusses, and W.T. Frankenberger Jr. 2000. Removal of Perchlorate in Ground Water with a Flow-Through Bioreactor. *Journal of Environmental Quality*, 29:578-583.

Gillogly, T., G. Lehman, L. Aldridge, D. Clifford, D. Roberts, and X. Lin. 2002. Ion Exchange Brine Treatment and Reuse for Perchlorate Contaminated Groundwaters. Presented at the 2002 Perchlorate Conference, Ontario, Calif., October 16-18, 2002.

Ghosh, A., D. Mishra, and W.P. Ela. 2002. Arsenic leaching from treatment residuals under landfill conditions, *Proceedings of the 2002 AWWA Annual Conference*. New Orleans, LA: AWWA.

Glass, C. and J. Silverstein. 1999. Denitrification of High Nitrate, High-Salinity Wastewater, Water research, 33(1):223-229.

Grant, G.T., E.R. Morris, D.A. Rees, P.J.C. Smith, and D. Thom. 1973. Biological Interactions Between Polysaccharides and Divalent Cations: The Egg Box Model. *FEBS LETT*. 32(1): 195-198.

Gingras, T.M. and J.R. Batista. 2002. Biological Reduction of Perchlorate in Ion exchange Regenerant Solutions Containing High Salinity and Ammonium Levels. *Journal of Environmental Monitoring*, 4:96-101.

Gu, B., G.M. Brown, P.V. Bonnesen, L. Liang, B.A. Moyer, R. Ober, S.D. Alexandratos. 2000. Development of a Novel Bifunctional Anion Exchange Resins with Improved Selectivity for Pertechnetate sorption from Contaminated Groundwater, *Environ. Sci. Technol.*, 34:1075-1080.

Gu, B., G.M. Brown, L. Maya, M.J. Lance, and B.A. Mayer. 2001. Regeneration of Perchlorate (ClO_4^-)-Loaded Anion Exchange Resins by a Novel Tetrachloroferrate ($FeCl_4^-$) Displacement Technique, *Environ. Sci., Technol.*, 35:3363-3368.

Guarini, W. 2002. Ex-Situ Biological Technologies for Removal of Perchlorate from Water, *Proceedings of Perchlorate and NDMA in Groundwater Conference*. Baldwin Park, CA.

Harper, T.R., and N.W. Kingham. 1992. Removal of Arsenic from Wastewater Using Chemical Precipitation Methods. *Wat. Environ. Res.* 64(3): 200-203.

Hart, E.J. and W.G. Brown. 1980. O(^{3}P) Atom Formation in γ- and 184.9-nm-Irradiated Aqueous Perchlorate Solutions, *J. Phys. Chem.*, 84, 2237-2240.

Hathaway, S.W. (deceased) and F Rubel, Jr. 1987. Removing Arsenic from Drinking Water. *JAWWA*, 78:8:61.

Haug, A., B. Larsen, and O. Smidsrod. 1974. Uronic Acid Sequence in Alginate from Different Sources. *Carbohydr. Res.* 32(2): 217-225.

Holt, J.G., N.R. Krieg, P. H. Sneath, J.T. Staley, and S.T. Williams. 1994. *Bergey's manual of Determinative Bacteriology – Ninth Edition*, Baltimore, Williams & Wilkins.

Hurley, J.M., W. Wallace, and E. Coppola. 1996. Ammonium Perchlorate Gradation. The Military Engineer. August-September.

Ionex. 2000. The ELIMINATE Process for Potassium Chloride Production, PuriTech bvba, Belgium.

Jackson, W.A., Rainwater, K., Anderson, T., Lehman, T., Tock, R., Mollhagen, T., and Moira Ridley. 2003. Distribution and Potential Sources of Perchlorate in the High Plains Region of Texas." Texas Commission on Environmental Quality Report.

Jang, L.K., G.G. Geesey, S.L. Lopez, S.L. Eastman, and P.L. Wichlaz. 1990. Use of Gel Forming Biopolymer Directly Dispensed into a Loop Fluidized Bed Reactor to Recover Dissolved Copper. *Wat. Res.* 24(7): 889-897.

Jing C., G.P. Korfiatis, and X. Meng. 2003. Immobilization Mechanisms of Arsenate in Iron Hydroxyde Sludge Stabilized with Cement. *Env. Sci. Tech.,* 37(21):5050-5056.

Kim, K. and M.D. Gurol. 1999. Investigation of Perchlorate Removal in Drinking Water Systems. *Proceedings of the AWWA National Conference*. Chicago, IL: AWWA.

Korenkov, V.N., V.I. Romanenko, S.I. Kuznetsov, and J.V. Voronov. 1976. Process for Purification of Industrial Waste Waters from Perchlorates and Chlorates. U.S. Patent 3,943,055.

Li, L. and E.N. Coppola. 2000. Final Report: Hydrothermal/Thermal Decomposition of Perchlorate, EPA Contract 68D99032.

Liang, S., J. Min, V. Snoeyink. 2000. Removal of Bromate and Perchlorate in Conventional Ozone/GAC Systems: Bench and Pilot Studies, *Proceedings of the AWWA Conference*, Denver, CO.

Logan, B.E., J. Wu, and F. Unz. 2001. Biological Perchlorate Reduction in High Salinity Solutions, *Water Research*, 35(12):3034-3038.

MacPhee, M.J., D.A. Cornwell, and R.N. Mutter. 2003. Disposal of Waste Resulting from Arsenic Removal Processes, AwwaRF, Denver, CO.

MacPhee, M.J., G.E. Charles, and D.A. Cornwell. 2001. Treatment of Arsenic Residuals from Drinking Water Processes, USEPA, Cincinnati, OH.

Martinsen, A., G. Skjak-Braek, and O. Smidsrod. 1989. Alginate as Immobilization Material: I. Correlation Between Chemical and Physical Alginate Gel Beads. Biotechnol. Bioeng., 33: 79-89.

Mayer, K. 2003. Overview on Perchlorate, Perchlorate Public Advisory Group Meeting. Los Angeles Regional Water Quality Control Board. Miller, J.P. and B.E. Logan. 2000. Sustained Perchlorate Degradation in an Autotrophic, Gas-Phase, Packed-Bed Bioreactor. Environ. Sci. Technol., 34:14:3018.

Metcalf and Eddy. 1991. Wastewater Engineering Treatment Disposal Reuse, 3rd ed. New York, N.Y.: McGraw-Hill Inc.

Miller, J.P. and Logan, B.E., 2000, Sustained Perchlorate Degradation in an Autotrophic, Gas-Phase, Packed-Bed Bioreactor, *Environ. Sci. Technol.*, 34(14):3018.

Min, J.H., and J.G. Hering. 1998a. Arsenate Sorption by Fe (III) Doped Alginate Gels. *Wat. Res.* 32(5): 1544-1552.

Min, J.H., and J.G. Hering. 1998b. Arsenate, Selenite and Chromate Sorption by Fe (III) Doped Alginate Gels. *Adv. Environ. Res.* 2(2): 207-217.

Min, J.H., and J.G. Hering. 2001. Calcium-Iron(III)-Doped Alginate Gel Beads for Sorption of Arsenate and Selenate, U.S. Patent No. 6,203,709.

Mofidi, A.A., J.H. Min, L.S. Palencia, B.M. Coffey, S. Linag, J.F. Green, . 2002. Advanced Oxidation Processes and UV Photolysis for Treatment of Drinking Water, Electrotechnology Applications for Potable Water Production And Protection of the Environment. California Energy Commission.

Narasimhan, R, C. Padilla, and L. Devkota. 2002. Design of a 1.5 MGD Arsenic Removal Plant for the City of Phoenix, *Proceedings of the AWWA National Conference.* New Orleans, LA: AWWA.

Nerenberg, R. and B.E. Rittman. 2001. Concurrent Perchlorate and Nitrate Reduction in a Novel, Hollow Fiber Membrane Biofilm Reactor. *Proceedings of the AWWA National Conference*. Washington, D.C.: AWWA.

Ogbonna, J. C., M. Matsumura, and H. Kataoka. 1991. Production of Glutamine by Micro-gel Bead-immobilized Corynebacterium Glutamicum 9703-T Cells in a Stirred Tank Reactor. Bioprocess Engineering. 7: 462-467.

Okeke, B.C., T. Giblin, and W.T. Frankenberger. 2002. Reduction of Perchlorate and Nitrate by Salt Tolerant Bacteria, *Environmental Pollution*, 118:357-363.

Office of Environmental Health Hazard Assessment. 2003. DRAFT Public Health Goal for Arsenic in Drinking Water, California Environmental Protection Agency Pesticide and Environmental Toxicology Section.

Parks, J.L., J. Novak, M. MacPhee, C. Itle, and M. Edwards. 2003. Effect of Ca on As Release from Ferric and Alum Residuals. *J. AWWA*, 95(6):108-118.

Rikken, G.B., A.G.M. Kroon, and C.G. van Ginkel. 1996. Transformation of (Per)chlorate into Chloride by a Newly Isolated Bacterium: Reduction and Dismutation. *Applied Microbiology Biotechnology*, 45:420.

Roberts, D., Y. Cang, X. Lin, D. Clifford. 2002. Biological Perchlorate and Nitrate Degradation in Spent Regenerant Ion Exchange Brine. Presented at the 2002 Perchlorate Conference, Ontario, Calif., October 16-18, 2002.

Romanenko, V.I., V.N. Korenkov, and S.I. Kuznetsov. 1976. Bacterial Decomposition of Ammonium Perchlorate. *Mikrobiologiya*. 61: 347-356.

Roote, D.S. 2001. Technology Status Report: Perchlorate Treatment Technologies, First Edition, TS-01-01, Groundwater Remediation Technologies Analysis Center.

SWRCB, 2001, California Ocean Plan, Water Quality Control Plan Ocean Waters of California.

Serp, D., M. Mueller, U. von Stocker, and I. W. Marison. 2002. Low- Temperature Electron Microscopy for the Study of Polysaccharide Ultrastructures in hydrogels. II. Effecte of Temperature on the Structure of Ca+2 – Alginate Beads. *Biotechnol. Bioeng*. 79(3): 253-259.

Schweiger, R.G. 1962. Acetylation of Alginic Acid. II. Reaction of Algin Acetate with Calcium and Other Divalent Ions. *J. Org. Chem*. 27(5): 1789-1791.

Tchobanoglous, G., and F.L. Burton. 1991. Metcalf and Eddy – Wastewater Engineering: Treatment, Disposal, and Reuse. New York: McGraw-Hill Inc.

USEPA. 1993. Office of Research and Development. Nitrogen Control Manual. September 1993.

USEPA. 1999. http://es.epa.gov/ncer/final/sbir/99/water/li99.html

USEPA. 2000. Regulations on the Disposal of Arsenic Residuals from Drinking Water Treatment Plants. (EPA/600/R-00/25)

Velings, N.M., and M.M. Mestdagh. 1995. Physicochemical Properties of Alginate Gel Beads. *Polym. Gels Networks*. 3(3): 311-330.

Venkatesh, K.R., S.M. Klara, D.L. Jennings, and N.J. Wagner. 2000. Removal and Destruction of Perchlorate and Other Anions from Ground Water Using ISEP+^{TM} System. In Perchlorate in the Environment, ed E. T. Urbansky, pp. 147–154. Plenum, New York.

Wallace, W., S. Beshear, D. Williams, S. Hospadar, and M. Owens. 1998. Perchlorate Reduction by a Mixed Culture in an Up-Flow Anaerobic Fixed Bed Reactor. *Journal of Industrial Microbiology and Biotechnology*. 20:126-131.

Wallace, W., T. Ward, A. Breen, and H. Attaway. 1996. Identification of an Anaerobic Bacterium which Reduces Perchlorate and Chlorate as *Wolinella succinogenes*, *J. Indust. Microbiol.*, 16:68-72.

Wang, L., A. Chen, and K. Fields. 2000. Arsenic Removal from Drinking Water by Ion Exchange and Activated Alumina Plants. EPA/600/R-00-088. Cincinnati, OH.

WHO. 2001. United Nations Synthesis Report on Arsenic in Drinking Water, Chapter 6: Safe Water Technology.

Woolard, C.R., and R.L. Irvine. 1995. Treatment of Hypersaline Wastewater in the Sequencing Batch Reactor, *Water Research*, 29(4):1159-1168.

Yakovlev, S.V., J.V. Voronov, V.N. Korenkov, A.B. Nevsky, V.A. Dobrikova, T.A. Karjukhina, I.N. Churbanova, and J.M. Laskov. 1973. Method for Biochemical Treatment of Industrial Waste Water. U.S. Patent 3,755,156.

Yamagiva, K., T. Kozawa, and A. Ohkawa. 1995. Effects of Alginate Composition and Gelling Conditions on Diffusional and Mechanical Properties of Calcium Alginate Gel Beads. *J. Chem. Eng. Japan*. 28(4): 462-467.

Yotsuyanagi, T., I. Yoshioka, N. Segi, and K. Ikeda. 1990. Effects of Monovalent Metal Ions and Propranol on the Calcium Association in Calcium Induced Alginate Gel Beads. *Chem. Pharm. Bull*. 28: 3124-3126.

Yotsuyanagi, T., I. Yoshioka, N. Segi, and K. Ikeda. 1991. Acid Induced and Calcium Induced gelation of Alginic Acid: Bead Formation and pH Dependent Swelling. *Chem. Pharm. Bull.* 39 (4): 1072-1074.

Zouboulis, A.I., and I.A. Katsoyuannis. 2002. Arsenic Removal Using Iron Oxide Loaded Alginate Beads. *Ind. Eng. Chem. Res.* 41(24): 6149-6155.

ABBREVIATIONS

AA	activated alumina
AL	action level
AWWA	American Water Works Association
AwwaRF	Awwa Research Foundation
ARA	Applied Research Associates
BDL	below detection limit
BOD	biochemical oxygen demand
°C	degrees Celsius
CDHS	California Department of Health Services
CCL	Contaminant Candidate List
CCR	Consumer Confidence Reports
CoD	Certificate of Destruction
CSTR	continuous stirred tank reactors
CTR	California Toxics Rule
DO	dissolved oxygen
DOC	dissolved organic carbon
DWTP	drinking water treatment plant
EBCT	empty bed contact time
EDR	Electrodialysis Reversal
EET	Environmental Engineering and Technology
EP	extraction procedure
EQSL	equivalent single length
°F	degrees Fahrenheit
ft	foot
ft^2	square feet
ft/s	foot per second
g	grams
G	guluronic
GAC	granular activated carbon
gal	gallon
gal/ft/d	gallon per foot per day
GFH	granular ferric hydroxide
gph	gallon per hour
gpm	gallon per minute
gpm/ft^2	gallon per minute per square foot
GWRS	Groundwater Replacement System
GWRTAC	Groundwater Remediation Technologies Analysis Center

h	hour
HDPE	high-density polyethylene
HFO	hydrous ferric oxide
HLR	hydraulic loading rates
HRT	hydraulic residence time
in.	inch
IX	ion exchange
ICP	inductively coupled plasma
Jour.	Journal
KDHE	Kansas Department of Health and the Environment
L	liter
L/h	liter per hour
LACSD	Los Angeles County Sanitation District
LADWP	Los Angeles Department of Water and Power
LC-MS	Liquid Chromatography-Mass Spectrometry
LPVCWD	La Puenta Valley County Water District
LSI	Langelier Saturation Index
M	mannuronic
MBR	Membrane Bio-reactor
MCL	maximum contaminant level
MCLG	maximum contaminant level goal
MF	membrane filtration
mg	milligram
m^2/g	square meter per gram
mgd	million gallons per day
mg/L	milligram per liter
mL	milliliter
MLSS	mixed liquor suspended solids
MRL	minimum reporting levels
µg/L	microgram per liter
µm	micrometer
NELAC	National Environmental Laboratory Accreditation Conference
NOAEL	No Observed Adverse Effect Level
NPDES	National Pollutant Discharge Elimination System
NTU	nephelometric turbidity unit
OCSD	Orange County Sanitation District
OEHHA	Office of Environmental Health Hazard Assessment
O&M	Operations and Maintenance
ORP	oxidation-reduction potential

PAC	Project Advisory Committee
PHGs	Public Health Goals
POTW	publicly owned treatment works
ppm	parts per million
psi	pounds per square inch
R	sand depth removal per scraping
R-1	first-stage reactor
R-2	second-stage reactor
RAS	Return activated sludge
RCRA	Resource Conservation Recovery Act
RfD	Reference Dose
RO	reverse osmosis
RoD	Record of Decision
rpm	revolutions per minute
SAWPA	Santa Ana Watershed Project Authority
SBR	sequencing batch reactor
Sec.	Section
SFBW	spent filter backwash water
STLC	soluble threshold limit concentration
SWRCB	State Water Resources Control Board (California)
TBLL	Technically Based Local Limit
TCLP	Toxicity Characteristics Leaching Procedure
TDS	total dissolved solids
THM	trihalomethane
TKN	Total Kjeldhal Nitrogen
TOC	total organic carbon
TS	total solids
TSS	total suspended solids
TTLC	total threshold limit concentration
TVS	total volatile solids
TVSS	total volatile suspended solids
UCMR	Unregulated Contaminant Monitoring Rule
U.S.	United States
USEPA	United States Environmental Protection Agency
VSS	volatile suspended solids
WAS	Waste activated sludge
WET	waste extraction test

IWA Publishing's authorised EU representative for General Product
Safety Regulations is Diane D'Arras, 15 rue Duret, 75116 Paris,
France, e-mail: safety@iwap.co.uk.

Printed and bound by CPI Group (UK) Ltd, Croydon, CR0 4YY
17/09/2026
02218813-0002